Providing Opportunities for the Mathematically Gifted, K–12

Peggy A. House, *Editor*

Contributors
James Bruni
Clare Heidema
Betty J. Krist
Charles E. Lamb
Beth Schlesinger

NATIONAL COUNCIL OF TEACHERS OF MATHEMATICS

Copyright © 1987 by
THE NATIONAL COUNCIL OF TEACHERS OF MATHEMATICS, INC.
1906 Association Drive, Reston, Virginia 22091
All rights reserved
ISBN 0-87353-239-2

The publications of the National Council of Teachers of Mathematics present a variety of viewpoints. The views expressed or implied in this publication, unless otherwise noted, should not be interpreted as official positions of the Council.

Printed in the United States of America

Contents

1. Overview and Description of Gifted Education 1

Historical Perspectives, 1
The Case for Gifted Education, 3
Who Are the Gifted? 4
Identifying Gifted Students, 9
The Psychology of Mathematical Giftedness, 14
Forces in the Lives of Gifted Children, 16
Organizational Alternatives for Gifted Programs, 21
Mathematical Content for the Gifted, 29
The Instructional Environment for the Gifted, 33
Evaluating Students and Programs, 36
Teachers for the Gifted, 41
Program Operation, 43

2. Guidelines 47

Sixteen Essential Components of Programs for the Gifted, 47
Elementary School Programs, 51
Middle, Junior, and Senior High School Programs, 59
Program Organization, 65
Initiating and Managing the Program, 69

3. A Closer Look 71

Special Schools, 71
Regional Center, 75
In-School Alternatives for Elementary Pupils, 77
Advanced Curricula for Secondary Schools, 80
Summer Programs, 81
Magnet Programs, 85
Mentor Programs, 86
Mathematics Contests, 87

4. Resources for Teachers of the Gifted 91

References 94

Appendix 99

Note to the Reader

A number of pages throughout the text feature problems that were given in a course for gifted high school students, followed by examples of the work they did in response (some of which also include the teacher's comments).

1
Overview and Description of Gifted Education

To claim that gifted children are our most valuable resource has become a cliche in American education; yet the record of our actions to translate that statement into practice has been sporadic and uneven. Rhetoric is almost always easier than action, but rhetoric alone does not solve problems. And in a time of national resurgence of rhetoric about all aspects of the educational system, what is most needed is action to implement programs of excellence.

But any educational reforms implemented in these closing years of the twentieth century must be designed in recognition of certain undeniable realities: First, our world will be increasingly permeated by technology, and mathematics and science will, therefore, be essential basic skills for all who would participate as contributing members of society. Second, we will need more than ever the special contributions of individuals with exceptional talent in these and other areas of achievement. Third, to nurture and cultivate such special talents requires deliberate effort through specialized programs and opportunities.

Historical Perspectives

This notion of concern for, and dedication to, the needs of the gifted and talented is by no means a new phenomenon. Ziv (1977, p. 3) recalled that even in ancient Greece, Plato called on society to nurture all especially bright children, no matter their social standing in life, and charged rulers "to scrutinize every child from birth . . . [and] to select all the children of gold, whether they come from parents of gold, or (as may occasionally happen) from parents of silver, iron or even bronze."

In our own century, Alfred North Whitehead warned, "In condition of modern life a rule is absolute: the race which does not value trained intelligence is doomed" (1931, p. 6). And Toffler (1970), projecting into the future society's need of people with high ability prepared to deal with the rapidly changing and complex world, argued that special efforts must be made to identify, train, and utilize those who are intellectually superior, or the world may face devastating consequences.

But although concern for the gifted has been expressed throughout history, the scientific study of giftedness is a twentieth-century phenomenon. Originally the term "gifted" was defined in terms of superior ability, which was commonly taken to mean intelligence, although neither the type nor the degree of superiority was clearly specified. With the development of IQ tests and the use of those tests by Terman in his studies of genius begun during the 1920s, the definition of giftedness in terms of intelligence was reinforced, an idea that predominated until recently when other conceptions of giftedness began to emerge, among them academic aptitude or potential, demonstrated academic ability, creative production, rate of development, talent, and mechanical or spatial abilities.

Not surprisingly, as the concept of giftedness evolved, so too did the research emphases of those who sought to examine the topic. At least three major approaches to the study of intellectual superiority can be identified: (1) the psychopathological approach, (2) the social impact approach, and (3) the statistical framework approach.

The psychopathological approach compares being gifted or talented to a mental disorder. From the 1890s until the early 1960s, genius was believed to be related to other attributes such as epilepsy, nervous abnormality, inferiority, or fixation of infantile desires. The social-impact approach looks at the talented and gifted in terms of their long-range contributions to society. Of course, a major disadvantage of this approach is that it severely limits identification of the intellectually superior, since it is not possible to identify them until long after their death. The statistical approach defines giftedness in terms of performance compared to the rest of the population. Ability is assessed by tests, with distribution following the normal curve. At the upper end of the curve, approximately three percent of the population is considered gifted or talented.

Despite the attention to the nature of giftedness and the needs of the gifted, several myths and misconceptions have arisen about the nature of being very bright and about special programs for the gifted. Among those that have been identified are the following beliefs:

- Programs for the gifted are undemocratic, and since bright children have special abilities, they don't need extra help.
- Special programs isolate the gifted from society.
- Resources are better used for other special groups.
- Gifted children always come from elitist backgrounds.
- Gifted students are socially maladjusted.

(Texas Education Agency 1979)

Misconceptions such as these are debilitating and impede the efforts of those who strive to improve the opportunities for maximum development of each individual's potential. In the pages that follow we will seek to dispel such myths.

The Case for Gifted Education

Societies traditionally place a high value on excellence, whether expressed through intellectual or artistic production, athletic prowess, technological or industrial leadership, wealth, military superiority, or some other societal value. Democratic societies, founded on the principles of equality and egalitarianism, at the same time profess strong belief in the values of individualism and personal freedom. Thus educators are propelled by the twin goals of equity in educational opportunity for all and the simultaneous provision for the particular needs of individuals. Yet equality and excellence frequently pull in opposing directions, and to date this country has made more progress toward the former goal than toward the latter. As Ziv (1977) pointed out, too often the educational system in the United States has supported gifted education only in times of crisis, such as in the post-Sputnik years. Tannenbaum (1979, p. 5) noted, "No other special group of children has been alternately embraced and repelled with no much vigor by educators and laymen alike." Gifted children have become a neglected minority in our schools. Recognizing this, the *Agenda for Action* (NCTM 1980, p. 18) warned:

The student most neglected, in terms of realizing full potential, is the gifted student of mathematics. Outstanding mathematical ability is a precious societal resource, sorely needed to maintain leadership in a technological world.

Perhaps we fear that fostering excellence in the gifted will compromise our commitment to the majority. To this end we weigh Clark's (1983) rationale for gifted education, which is based on the following arguments:

1. In order to retain giftedness as well as to develop potential, children need to participate in special programs.
2. As Thomas Jefferson noted, there is nothing more unequal than the equal treatment of unequal people.
3. Restricting development can lead to boredom, anger, and frustration.
4. Providing for the needs of all is not elitism.
5. Gifted students need programs that do not allow them to become isolated members of society.
6. Good programs foster development and unleash learning potential.
7. Contributions to society resulting from the efforts of gifted students traditionally occur at a rate that exceeds the proportion of the gifted in the population.

It is hard to imagine that any educator, aware that a learning disabled child was experiencing difficulties in school, would deliberately neglect that individual or refuse to extend support and encouragement whenever possible. Yet the failure of educators to respond to the comparable needs of the gifted can be seen from the following list of what Galbraith (1983, 1984) identified as the "great gripes of gifted kids":

- No one explains what being gifted is all about; it's kept a big secret.
- The stuff we do in school is too easy and it's boring.
- Lots of our coursework is irrelevant.
- We feel too different and wish people would accept us for what we are.
- Parents, teachers, friends expect us to be perfect all the time.
- Friends who really understand us are few and far between.
- Peers often tease us about being smart.
- We feel overwhelmed by the number of things we can do in life.
- We worry a lot about world problems and feel helpless to do anything about them.

The implications are clear: Special children have special needs, and the needs of gifted students are not well served by our educational system. We have not yet fulfilled our responsibility, either to the gifted individuals or to society, to translate into practice not only the principles of equality but also the ideals of excellence for all. Let us summarize the major issues that must be addressed by educators who hope to provide opportunities for the mathematically gifted, K–12.

Who Are the Gifted?

The two issues most discussed in the literature on gifted education are the definition of giftedness and the identification of the gifted. Of the two, the most fundamental is the question of definition, for without a clear operational definition of the target population, identification will surely flounder.

Thus, the starting point in providing for the needs of the gifted and talented must be to formulate a definition of who constitutes the population in question. Perhaps because answers to the question of what makes giftedness reflect the prevailing values and concerns of society, defining giftedness has been a continuing issue in education. Nevertheless, establishing a clear, workable definition of the gifted is a prerequisite to determining identification procedures and programs.

Prior to about 1950, most educators and school systems relied on IQ tests and scholastic achievement scores to define giftedness, such as in the following definition by Terman et al. (1926):

[Giftedness is] the top one percent level in general intellectual ability, as measured by the Stanford-Binet Intelligence Scale or a comparable instrument.

The late 1950s and early 1960s brought increased emphasis on intellectual superiority, and attention turned to other aspects of giftedness. Indeed, the concern of many researchers was how to view intelligence. Guilford (1959, 1967), in his structure of the intellect model, examined other facets of the individual apart from IQ. Other researchers, like Getzels and Jackson (1958, 1962) and Torrance (1965), contributed significantly by adding creativity to the characterization of giftedness.

In 1972, a report to Congress from the United States Office of Education (USOE) contained a definition of giftedness that has been widely adopted by school systems and states. According to former Commissioner of Education Marland (1972, p. 2):

Gifted and talented children are those identified by professionally qualified persons who by virtue of outstanding abilities are capable of high performance. These are children who require differential educational programs and/or services beyond those provided by the regular school program in order to realize their contribution to self and the society.

Children capable of high performance include those with demonstrated achievement and/or potential ability in any of the following areas, singly or in combination:

1. General intellectual ability
2. Specific academic aptitude
3. Creative or productive thinking
4. Leadership ability
5. Visual and performing arts
6. Psychomotor ability

(It should be noted that in 1978 psychomotor ability was dropped from the definition above because it was assumed that this area was well served by extracurricular opportunities in schools.)

Acknowledging that the USOE definition served a useful purpose in calling attention to a wider range of abilities, Renzulli (1978) nevertheless criticized it on three counts: (1) its failure to include nonintellective (motivational) factors; (2) the nonparallel nature of its six categories, two of which (2 and 5) refer to general performance areas in which talents are manifested and of which the remaining four denote processes that may be brought to bear on a specific aptitude; and (3) its tendency to be misinterpreted and misused by practitioners. The misuses identified by Renzulli were the tendency to treat the six categories as though they were mutually exclusive and the tendency to give lip service to the definition while continuing to use intelligence or aptitude test scores as criteria for entrance into special programs.

In an effort to remove the deficiencies he perceived in the USOE definition, Renzulli proposed instead a three-ring model in which giftedness was defined as the intersection of three overlapping clusters of traits: above-average general ability, task commitment, and creativity (fig. 1). In setting

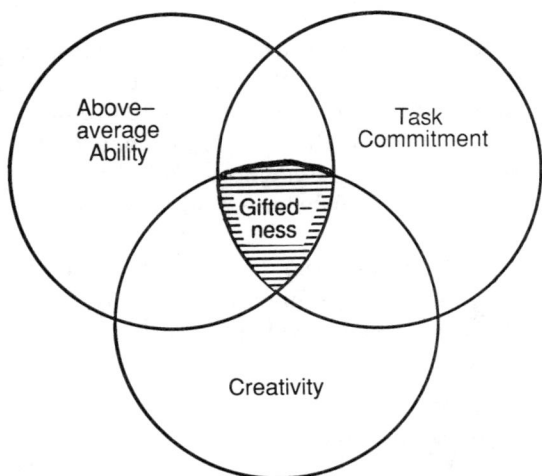

Fig. 1

forth his model, Renzulli was explicit in stating that no one of these clusters alone makes giftedness, and that each is an equal partner in contributing to the phenomenon. Thus, Renzulli concluded (1978, p. 261):

> Giftedness consists of an interaction among three basic clusters of human traits—these clusters being above average general abilities, high levels of task commitment, and high levels of creativity. Gifted and talented children are those possessing or capable of developing this composite set of traits and applying them to any potentially valuable area of human performance. Children who manifest or are capable of developing an interaction among the three clusters require a wide variety of educational opportunities and services that are not ordinarily provided through regular instructional programs.

Although these approaches to defining giftedness are among the best known, other models have been proposed as well. Some persons, for example, define giftedness in terms of a specific percentage of the population on some trait. Legislation in support of gifted education often is written in such terms. Another less-known model is that of Cohn (1981), in which giftedness is defined as consisting of four ability domains: the intellectual domain (quantitative, verbal, spatial talent); the artistic domain (fine arts, performing arts); the social domain (leadership, empathic or altruistic talent); and other human ability domains. Other models and definitions can be found as well.

The important message for educators responsible for programs for the gifted is that a defensible operational definition of giftedness is essential for at least two reasons: first, it becomes the basis of determining identification and selection procedures, and second, it gives direction to the development of educational programs and opportunities. The definitions that educators formulate will vary from one situation to another, but they should reflect careful consideration of accepted educational principles, previous research, and community values. In addition, it should be recognized that definitions may range from restrictive or conservative to inclusive or liberal.

Conservative definitions, such as those stated in terms of a percentage of the population, may be useful for some purposes such as determining how to allocate monies. For example, state funds may be distributed according to a formula based on a specified percentage of the school population and the assumption of random distribution of talent. Or a definition might limit participation to performance in a specific domain, such as a score in the top percentile on a mathematics aptitude or achievement test without regard for verbal or spatial ability.

Liberal definitions, on the other hand, broaden the concept of giftedness to include a wider range of human activity. Some, for example, attempt to include in their definitions any individual whose performance in an area of human activity is consistently superior.

Several consequences arise from the degree of restrictiveness or inclusiveness of the definition. On the one hand, restrictive definitions lend themselves to tighter identification procedures, but they may also lead to the unintended exclusion of some students. Liberal definitions, on the other hand, may include more individuals, but they usually rely on less precise measures of performance and on more subjective judgments. They also tend to produce more diverse groups of participants, an outcome that has implications for the programming decisions that must be made.

The goal of specifying a concise operational definition is, in reality, often confounded by the use of such ill-defined terms as ability, aptitude, intelligence, achievement, potential, creativity, or productivity. Such terms often are heavily laden with hidden meanings, values, or policy concerns. In addition, there is in the literature a growing list of labels that may influence the formulation of a definition as well as the establishment of an identification process. Examples are "creatively gifted," "culturally different gifted," "gifted underachiever," "disadvantaged gifted," and "latently gifted."

Perhaps to lessen the abstract nature of constructs like those above, definitions of giftedness frequently are augmented by lists of characteristics of gifted children. Such lists, which can be helpful for purposes of identification, should be viewed as examples of commonly observed traits or indicators, not as exhaustive or exclusive, and they should suggest the multifaceted nature of giftedness. Gifted children as a group usually exhibit more variability than do average children as a group.

The definition of giftedness adopted for a program should direct the identification and selection processes, not the reverse. Therefore, educators must develop and critique their definitions in terms of how useful, reliable, and effective they are for identification. Whenever a child is or is not identified for a gifted program, that identification should be defensible on the basis of the definition, not the procedures employed in the identification process. Here it is also useful to differentiate between identification and selection. Limitations of financial or personnel resources may, in reality, place constraints on the number of students who can be selected for par-

Characteristics of Giftedness

Any list of characteristics for gifted and talented children should be viewed as examples of possible traits or indicators that such children may have. Few children will exhibit all the attributes that are listed.

General Behavior

- Early, avid reader with good comprehension and large vocabulary
- Early memory for verses, songs, stories, etc.
- Quick mastery of basic skills
- Advanced spatial ability
- Organizer and leader; able to manipulate or influence others
- Early sense of justice; concern for fairness
- Perceptive in interpreting both verbal and nonverbal cues
- Ability to construct and handle abstractions
- Ability to concentrate and work independently for longer periods of time
- Self-starter; self-directed; strives for perfection
- Active listener; highly inquisitive
- Interests are both focused and eclectic
- Enjoyment of new things and new ways of doing things
- Well organized and efficient
- Energetic; enthusiastic about new ideas or challenges
- Good sense of humor

Learning Behavior

- Pleasure in intellectual activity
- Keen power of observation; eye for the important
- Ability to abstract, conceptualize, and synthesize
- Insight into cause/effect relationships
- Questioning attitude, seeking information for its own sake and using a wide variety of resources
- Skeptical, critical, and evaluative
- Large knowledge base and recall ability
- Ability to grasp underlying principles and to make generalizations
- Perceptive of similarities, differences, and anomalies
- Ability to convey ideas effectively

Creative Behavior

- Fluent thinker; able to see many possibilities and consequences
- Flexible thinker; able to use alternative approaches
- Original thinker; able to see relationships

- Elaborative thinker; able to find new responses
- Good guesser and hypothesizer
- High level of curiosity
- Intellectually playful and imaginative
- Intellectually uninhibited and inventive
- Sensitive to the aesthetic dimensions
- Impulsive and emotionally sensitive
- Often bored by routine tasks

Mathematical Behavior

- Early curiosity and understanding about the quantitative aspects of things
- Ability to think logically and symbolically about quantitative and spatial relationships
- Ability to perceive and generalize about mathematical patterns, structures, relations, and operations
- Ability to reason analytically, deductively, and inductively
- Ability to abbreviate mathematical reasoning and to find rational, economical solutions
- Flexibility and reversibility of mental processes in mathematical activity
- Ability to remember mathematical symbols, relationships, proofs, methods of solution, etc.
- Ability to transfer learning to novel situations
- Energy and persistence in solving mathematics problems
- Mathematical perception of the world

ticipation in a program, but identification of eligible or potentially eligible candidates should continue to rest on the definition.

In general usage, the terms *gifted* and *talented* are often equated. Dictionaries commonly list the two terms as synonyms and define "gifted" as "having natural talent." In the scientific literature, a few authors have attempted to distinguish between the two, usually applying the term *gifted* to a more general category and using *talented* to describe exceptional ability in a specified area. More often, however, definitions of giftedness encompass the concept of talent as well. Unless there is some programmatic reason to differentiate between the two, there is probably little to be gained by insisting on clearly distinguishable definitions. In this publication, the terms *gifted* and *talented* are used interchangeably.

Identifying Gifted Students

Identification, along with the concern over definition, has dominated the literature on gifted education; the approaches to identification, like those to definition, have changed dramatically during the past two or three decades.

During the 1950s and 1960s researchers began concentrating on facets of giftedness other than intelligence and achievement, and they began to seek more equitable ways to identify the gifted in the humanistic setting that prevailed at that time. With a renewed interest in the gifted and a broadened definition of giftedness sparked by the Marland report of 1972, states and schools again began to seek approaches other than IQ and achievement testing as the means to identification. Nevertheless, even today many schools continue to rely solely on standardized tests, grades, and teacher nominations to select students for special programs.

The choice of any broadened definition of giftedness must generate not reliance on a single score but a more complex identification process leading to selection of students for the alternative programs. Feldhusen et al. (1984) suggested a five-step approach to a sound identification process. What follows is an elaboration of considerations surrounding an identification scheme based on that approach.

Step 1: Define the goals of the program and the types of students to be served by it. This step is naturally tied to the definition process, since the program's goals may serve some gifted students but not others. These program goals, in turn, should be determined after an assessment of the needs of the gifted and talented students in a particular setting. Care must be taken to assure that assessment procedures do not overlook disadvantaged or underachieving students, and the goals that are formulated should realistically reflect the resources available to the program. Defensible identification procedures are based on clear, defensible goals.

Step 2: Decide on nomination procedures that will find all qualified candidates for the program. In order not to overlook some potentially qualified students, all available sources of information should be used in the nomination process. Examples of possible sources include school records that include developmental information; test scores and other performance information, including anecdotal evidence and samples of creative work; behavior checklists and rating scales; self-assessments and peer ratings; and nominations from teachers, parents, and other persons familiar with the students' work or qualified to judge their involvement in activities outside of school.

Any source of information does, of course, have limitations as well as advantages, and these should be recognized as such. School records, for example, suffer from being highly variable, often incomplete, uncertain as to the accuracy of some information, and more often judgmental than descriptive of behavior. Yet, despite these inadequacies, they can contribute valuable information for assessing an individual's academic and cognitive development. Further, they may contain samples of work or descriptions of accomplishments.

Behavior rating scales are based on characteristics of gifted students and rely on direct observation rather than on inferences drawn from test scores.

Many believe that such scales are effective in recognizing individuals who might otherwise be overlooked. In using such instruments, program directors should take care both to verify the reliability and validity of the scales and to provide adequate training for those persons who will conduct the ratings.

There is in the literature considerable evidence to indicate that teachers fail to nominate many gifted students, usually because the gifted frequently violate the classroom expectations that teachers hold for "good students." In particular, teachers overlook underachievers, culturally different children, pupils with motivational or emotional problems, children with negative attitudes, children who do not conform to classroom norms of behavior, children with poor study habits, and children who think in original or "nonstandard" ways. Indeed, several studies have revealed that teachers failed to nominate over half of their gifted pupils. In one study, Jacobs (1971) found that although primary school teachers identified fewer than 10 percent of the gifted children in their classes, parents correctly selected 61 percent.

Step 3: Decide on the assessment procedures that will appropriately screen the nominees for the proposed program. As in the nomination process, multiple instruments should be used for the assessment procedure as well. At this stage, assessment should be viewed not as "weeding out" the unqualified but as a means to provide information on which to base individual programming decisions. Unless the program goals are very narrow, the students to be served will undoubtedly have a variety of strengths and needs that must be taken into account when planning programs. The more information that is available, the more effective will be the programs offered.

A particular problem exists in the case of disadvantaged or culturally different gifted students since some instruments are biased against these groups. In reporting the findings of a survey of over 200 school districts, Alvino et al. (1981) described widespread misuse and abuse of tests to identify gifted students among cultural subgroups. Commenting on those findings, Rogers (1986, p. 2) noted, "It became apparent that national identification practices reflected flagrant use of tests and instruments with populations on which they were not normed and for which they were never intended."

A large number of tests and rating scales are available, some more appropriate than others for identifying the gifted. These include individual and group intelligence tests, achievement tests, aptitude tests, creativity tests, and so on.

Group IQ and achievement tests, although popular, have serious limitations. They are designed for average students, and consequently the ceilings on these tests may be too low to discriminate between the bright and the truly gifted. A common criticism is that frequently such tests predominantly measure factual recall and low-level thinking skills. Because of the objective

2. Find the one millionth term of the following sequence:

   1, 2, 2, 3, 3, 3, 4, 4, 4, 4, 5, 5, 5, 5, 5, . . .

   (Describe fully the method you used to solve this problem. Your solution will be evaluated for completeness, accuracy, clarity, organization, and originality.)

The pattern is one 1, two 2's, three 3's, and so on. Call each repetition a cycle. The number of terms encountered after each cycle runs as follows:

Cycle	Term ated
1	1
2	3
3	6
4	10
5	15
etc.	

The numbers at right are triangular numbers, with the formula $\frac{n(n+1)}{2}$.

It is necessary to find the lowest number n such that $f(n) \geq 1,000,000$ as this number would be the millionth term.

Solving for n: $\frac{n(n+1)}{2} \geq 1,000,000 \rightarrow n(n+1) \geq 2,000,000$

For $n =$ 1,413: $\frac{1,413(1,414)}{2} = 998,991$

1,414: $\frac{1,414(1,415)}{2} = 1,000,405$

Thus the millionth term will be reached during the 1,414th cycle. The millionth term is $\boxed{1,414}$.

---

the sequence
1
2 2
3 3 3
4 4 4 4
5 5 5 5 5

row completed
1
2
3
4
5

Let $n =$ the number of rows completed to get 10^6 items in each row; there are n items in each row; $1+2+3+\ldots +n = 10^6$

$S = \frac{n(n+1)}{2}$

$10^6 = \frac{n(n+1)}{2}$

$2 \cdot 10^6 = n^2 + n$

$n^2 + n - 2,000,000 = 0$

$n = \frac{-1 \pm \sqrt{1 + 8,000,000}}{2}$

$n = \frac{-1 \pm 2827.4873}{2}$

$n = 1413.71365$

This shows that 1413 rows have been completed, and about 3/4 of the 1414th row has been completed.

1414 is the millionth term.

good — nice presentation

format of group tests, they generally do not allow for the divergent or insightful thinking that characterizes gifted children. Further, their reliance on the printed word means that they may fail to recognize gifted students among those with reading difficulties, cultural diversity, records of underachievement, or motivational or emotional problems.

Because of the limitations inherent in group tests, most educators recommend individual intelligence tests as the best method for identifying the gifted. However, these are costly in both time and money, and they require the services of a trained psychologist. Consequently they are impractical as a general screening tool in many schools.

Complex qualities like creativity and leadership are difficult to define and correspondingly difficult to measure. Creativity tests are newer than IQ or achievement tests, but some show promise for identifying divergent thinkers who may be overlooked or even penalized by those more established tests. As with other instruments, creativity tests should be carefully examined for validity and reliability; most require trained administrators or evaluators. In general, these tests should be used in conjunction with other measures of intelligence and achievement.

Biographical inventories are also valuable for providing information about students' interests and background. In this case, specific items from the inventory may be used either separately or in conjunction with other items, a freedom not possible with standardized tests. This enables more direct screening for behavior related to the program. The Institute for Behavioral Research in Creativity (1974) found, both from a survey of the literature and from their own study, that biographical inventories were generally effective as predictors of potential academic talent.

Assistance in locating and selecting instruments for identification can be found in the report by Richert et al. (1982), which includes a list of over sixty tests, rating scales, checklists, and inventories rated by a national panel of experts for their applicability to (*a*) all categories of giftedness, (*b*) advantaged versus disadvantaged populations, (*c*) various age levels, and (*d*) stage of the identification process.

Step 4: Differentiate individuals by examining all the identification data and student profiles. This step involves coding and summarizing a large amount of data into a profile or case study of each student. It may include interviews with students, parents, or teachers in an effort to better understand the information.

There is a danger in trying to "sum up" these data in some way that yields a composite score, since the result may obscure valuable indicators of potential. Further, scores from different types of instruments represent diverse characteristics, and it is statistically unsound to attempt combining them. Perhaps the most important outcome of the case studies is the identification of individuals who demonstrate giftedness in alternative ways.

Step 5: Provide for evaluation of the identification process. The final step involves collecting evidence to assure that the identification procedures reflect the goals of the program and that they are effective in achieving those goals. Correlations of identification instruments with criteria for success in the program should be used in reassessing choices of appropriate instruments. It is a good idea to include in the evaluation both students who were identified for the program and those who were not.

From the identification process flows the selection of participants for the program. Ideally, all those identified for the program should also be eligible for selection. However, limited resources often place constraints here that lead to the necessity of making difficult choices. At the same time, a strong, defensible identification process that locates more gifted students than current programs can accommodate provides solid data on which to base one's case for expanded resources.

The Psychology of Mathematical Giftedness

Educators cannot help the mathematically gifted realize the full potential of their outstanding abilities unless they understand the characteristics and needs of their students. Much of what we know about those characteristics are the results of the observations conducted by the Soviet psychologist V. A. Krutetskii as reported in *The Psychology of Mathematical Abilities in Schoolchildren* (1976).

Krutetskii concluded that the mathematically gifted have a unique neurological organization that results in a "mathematical cast of mind." This trait often emerges in elementary form by age seven or eight and later acquires a very broad character. It is expressed in a striving to make the environment mathematical; in a constant urge to pay attention to the mathematical aspects of phenomena; in noticing spatial and quantitative relationships, bonds, and functional dependencies everywhere—in short, to see the world "through mathematical eyes" (Krutetskii 1976, p. 302).

The mathematically gifted are capable of logical thought; they deal in abstractions; they think in mathematical symbols. They can reason swiftly and solve unfamiliar problems rapidly. As Heid (1983, p. 222) described it, "The logic used by the gifted is so readily applied that it seems to be not so much a learned ability as an almost innate characteristic of their cognitive processing."

Mathematically gifted individuals are capable of rapid and broad generalization of mathematical ideas, and unlike their less able peers, they are more focused on the underlying relationships and general structure of problems than on the specifics of irrelevant detail. "It is not uncommon for a gifted student to solve a problem on its most general level, to generalize algorithms for solving whole categories of problems of the type given, and then to neglect answering the particular question stated in the problem" (Heid 1983, p. 223).

The gifted think flexibly and possess economy of thought. As they mature, they develop the need to find the clearest, most logical, most elegant solution to a problem. They will search for alternative solutions to problems if they are not totally satisfied with their initial results.

They are able to abbreviate or curtail the process of mathematical reasoning as well as to reverse their thought processes. For example, after proving a theorem, they can then rapidly prove its converse, a tendency not found in the nongifted.

The gifted have a mathematical memory and easily retain mathematical material, relationships, proofs, or methods of solution, even over long periods of time. They do not tire when doing mathematics, a characteristic observed even in young gifted children.

Krutetskii's work also revealed the existence of distinguishable types of mathematical casts of mind among gifted students. Even among pupils whose mathematical experiences were limited to school mathematics, three groups were identifiable: (1) analytic types, (2) geometric types, and (3) harmonic types (Krutetskii 1976, pp. 315–29).

Analytic thinkers possess a mathematically abstract cast of mind. In their thinking a well-developed verbal-logical component predominates over a weak visual-pictorial one. They function easily with abstract patterns and show no need for visual supports when considering mathematical relationships. They will, in fact, employ complicated analytical methods to attack problems even when visual approaches would yield much simpler solutions. They prefer abstract situations and will attempt to translate concrete problems into abstract terms whenever possible. They may have weakly developed spatial visualization abilities, especially for three-dimensional relationships. In school they are more likely to excel in arithmetic and algebra than in geometry.

Geometric thinkers exhibit a mathematically pictorial cast of mind. Their thinking is driven by a well-developed visual component that impels them to interpret visually expressions of abstract mathematical relationships, sometimes in very ingenious ways. Although their verbal-logical abilities may be quite well developed, they persist in trying to operate with visual schemes even when a problem is readily solved by analytic means and the use of visual images is superfluous or difficult. Indeed, these students frequently find that functional relationships and analytical formulas become understandable and convincing only when given a visual interpretation.

Harmonic thinkers exhibit a relative equilibrium between the extremes of the other two types. They possess both well-developed verbal-logical and well-developed visual-pictorial abilities, and when given a problem, they are usually capable of producing solutions of both kinds. Krutetskii observed two subtypes among harmonic thinkers: those with an inclination for mental operations without the use of visual means and those with an inclination for mental operations with the use of visual means. In other words, although harmonic thinkers are perfectly capable of representing relationships pictorially, some prefer to do so while others see no need for it.

In summary, we can identify from Krutetskii's work the following significant traits of the mathematically gifted (1976, pp.350–51):

- Formalized perception of mathematical material and grasp of the formal structure of problems
- Logical thought about quantitative and spatial relationships and the ability to think in mathematical symbols
- Rapid and broad generalization of mathematical objects, relations and operations
- Curtailment of mathematical reasoning and the ability to think in curtailed structures
- Flexibility of mental processes
- Striving for clarity, simplicity, economy and rationality of solutions
- Rapid and free reconstruction of a mental process as well as reversibility of mathematical reasoning
- Generalized memory for mathematical relationships, characteristics, arguments, proofs, methods of solution, and principles of problem solving
- A mathematical cast of mind
- Energy and persistence in solving problems

Forces in the Lives of Gifted Children

There is a very real danger that, critical as they are, the questions of defining and identifying giftedness may cloud educators' perceptions to the extent that they fail to consider a complex of very substantial but nebulous factors that impinge upon the gifted. Among the forces that interact in the lives of gifted children and the world in which they live are personal concerns of the students themselves; parents' expectations, peer pressure, administrative constraints, equity issues, and cultural, ethnic, and sex differences.

Students' Personal Concerns

Perhaps the most frustrating aspect of being gifted is not understanding what giftedness is all about. Gifted children may know they are different, but they do not comprehend either the nature or the significance of their differences. They require adult assistance in realizing that different does not mean "weird," that they need not hide their talents, and that it is "okay" not to be gifted in everything. Sooner or later, gifted children are likely to encounter individuals who resent their talents and who begrudge them their special opportunities on the grounds that the gifted need no extra advantage. Children require help to learn to deal with such prejudices.

Gifted children frequently experience an acute fear of failure, and others must realize that "failure" to a gifted student may mean something very different than it does to an average child, such as getting a B+ instead of

8. Find the sum of all the positive proper fractions with denominators less than or equal to 100:

$$\frac{1}{2} + \frac{1}{3} + \frac{2}{3} + \frac{1}{4} + \frac{2}{4} + \frac{3}{4} + \ldots + \frac{1}{100} + \frac{2}{100} + \ldots + \frac{99}{100}$$

$$\underbrace{\frac{1}{2}}_{\frac{1}{2}} + \underbrace{\frac{1}{3} + \frac{2}{3}}_{1} + \underbrace{\frac{1}{4} + \frac{2}{4} + \frac{3}{4}}_{1\frac{1}{2}} + \underbrace{\frac{1}{5} + \frac{2}{5} + \frac{3}{5} + \frac{4}{5}}_{2} \ldots$$

The sum of all fractions with the same denominator is the denominator minus 1 multiplied by $\frac{1}{2}$.   Ex: $\frac{1}{2}(5-1) = \frac{1}{2} \cdot 4 = 2$

If 100 is the highest denom. $\frac{1}{2}(100-1) = 49.5$

$$\underbrace{\frac{1}{2} + 1 + 1\frac{1}{2} + \ldots 48\frac{1}{2} + 49 + 49\frac{1}{2}}_{49+1 = 50}$$

$\frac{1}{2} n (50) =$ sum $n =$ the number of denom.

$\frac{1}{2} \cdot 99 (50) = \boxed{2475}$

The series given is:

$$\frac{1}{2} + \frac{1}{3} + \frac{2}{3} + \frac{1}{4} + \frac{2}{4} + \frac{3}{4} + \frac{1}{5} + \frac{2}{5} + \frac{3}{5} + \frac{4}{5} + \ldots + \frac{1}{100} + \frac{2}{100} + \ldots + \frac{99}{100} = S$$

Group like denoms.
$$\frac{1}{2} + \frac{3}{3} + \frac{6}{4} + \frac{10}{5} + \ldots + \frac{4950}{100} = S$$

Make each denominator "2"
$$\frac{1}{2} + \frac{2}{2} + \frac{3}{2} + \frac{4}{2} + \ldots + \frac{99}{2} = S$$

Can be reduced to:
$$\frac{1 + 2 + 3 + 4 + \ldots + 99}{2} = S$$

$$\frac{4950}{2} = S$$

$$2475 = S$$

∴ the sum of all the fractions in the given series is 2475

an A grade. Bright students have no difficulty excelling in a regular class, but when placed in a special program, they may not cope well with no longer being the "best." Some gifted students have refused invitations to honors classes for fear that their grades might be lower than the sure A from the regular class. In this regard, it is especially critical that gifted children realize that their value and self-worth do not arise solely from the work they produce or the level of their success with every task.

At the same time, gifted students are frequently bored and impatient with school and with what they perceive as needless repetition of trivial material. When asked, for example, to complete the sentence, "In my mathematics class I spend most of my time . . ." several hundred gifted high school students overwhelmingly replied in one of three ways: listening to the teacher talk, working alone on assignments, or engaging in some unrelated or unproductive behavior out of boredom with what is going on in class. These students have a right to experience education as a relevant, challenging, and engaging enterprise.

Parental Involvement

Parental involvement in the education of their gifted children is of crucial importance, but parents can be either a great source of help or a great headache to educators, depending on how well they understand what is happening to their children (Malone 1975). On the one hand, parents may expect too much of their high-ability children and thus place unreasonable pressure on them to participate and excel in programs for the gifted. Frequently, too, parents expect equal levels of excellence in all school subjects and in other areas of accomplishment. Children thus may come to believe that their parents' love and acceptance are dependent on the child's performance. On the other hand, parents can be strong allies when they encourage their children to learn and grow in their own unique styles.

Because it is a fact frequently overlooked, it also should be pointed out that there is a prevailing stereotype that gifted children come from middle and upper class families with solid family structure and more than adequate material resources. As will be noted later, gifted children from other socioeconomic backgrounds frequently go unidentified and unserved. Further, gifted children are not immune to family pressures, including financial hardships, sibling rivalry, divorce, birth and death, and a multitude of other physical and psychological pressures that can, at any time, impede their performance in and out of school.

Peer Pressure

As children grow up, peer pressure becomes an ever-increasing force in their decision-making processes. Many gifted children, in particular, experience negative peer response in the form of ridicule and misunderstanding. For these children peer pressure sometimes presents a no-win situation: When they succeed, classmates tease them for being smart; when they don't succeed, they tease them for failing. The gifted also complain that their friends don't understand them and don't share their interests. Frequently, too, gifted children seek the companionship of older youngsters who are more nearly their intellectual peers.

Programs for the gifted must strike a balance between providing an environment where bright students can be stimulated by interaction with their intellectual peers while at the same time helping to socialize them to function effectively with children who are less gifted than they.

The School Environment

Educational programs for the talented will be surrounded by a web of administrative constraints. One is that school administrators and teachers understandably want to retain gifted students in the local school, but as a result they may stand in the way of enabling qualified students to participate in programs for the gifted. Among the forms that this resistance can take are failure to nominate qualified students for talent searches, refusal to accept credits earned outside the home school, refusal to arrange special schedules for the gifted, and insistence that gifted students must take courses in lockstep even when they have already demonstrated mastery of the material of a given grade or course.

Other hurdles that may have to be jumped deal with adjusting schedules, especially when that involves sending a gifted child to another school (such as a junior high school pupil to the senior high school), transportation to special programs, and provision of special materials and resources. In some situations, programs for the gifted are contingent on grant proposals, extra reports, and separate applications for special funds that demand time and allocations of personnel. Each local system will have its own unique set of constraints, but persons responsible for developing and implementing programs for the gifted cannot afford to ignore these problems or to treat them lightly.

Equity Issues

All special programs must recognize and be concerned about equity of access and opportunity. Issues that must be faced squarely include these: Is it fair to put resources into programs for superior students when there is great need for programs for below-average students? Do special programs serve only children from advantaged backgrounds? Will special programs favor children from a particular race or ethnic background? Will special programs favor one sex over the other? To what extent are special programs dependent on the parents' ability to pay?

Equity can be a two-edged sword. On the one hand, public education is founded on the principle of equal access for all cultural and ethnic groups. On the other hand, a program that succeeds in achieving cultural pluralism may face new problems. In some cultures, for example, females are not expected to perform on a par with males. Some cultures or family traditions may not value educational excellence, and students who do very well may not receive support from the family. Children from certain traditions are expected to be self-effacing and nonaggressive about applying for special opportunities. One culture may value conformity and compliance rather than independent thinking; another may stress individual performance and competitiveness rather than cooperation; either case poses conflicts with the strategies likely to be encountered in a gifted education program. Still other cultures place total responsibility for education on the school system, and family support will not be forthcoming. Minority populations, in particular, are likely to experience such realities.

Evidence from a variety of studies also indicates that by the time they reach secondary school, males outperform females in mathematics. For example, many talent searches have found that although girls' representation in the talent search is close to their proportion in the population, they are not so represented among those finally selected for the programs. Mathematical competitions also have been dominated by boys. And girls are far more likely than boys to drop out of special programs for the mathematically talented. Other indicators include data on tests scores, courses taken, careers chosen, spatial visualization abilities, and the like. Dissenters counter by attributing the differences to factors such as environment, social discrimination, and a complex variety of genetic, physiological, and emotional factors. Whatever the eventual outcome of these debates, the fact remains that the track record of girls in programs for the mathematically gifted has not been comparable to that of boys.

Underachievement

Clark (1983, p. 323) calls underachievement by gifted students "one of the most baffling, most frustrating problems a parent or teacher can face." This underachievement either can be situational, such as when a home problem arises or the student has a conflict with a particular teacher, or may be chronic and highly resistant to treatment. If we discount cases caused by physical illness or disabilities, underachievement seems to result from a complex interaction of internal personality factors and external sources including societal pressures, family expectations, and the school environment.

"Gifted underachievers" are students who are clearly acknowledged as gifted but who do not perform at the levels expected by parents and teachers. Their underachievement may occur in a given subject area only, or it may affect the entire academic program. Underachievers usually exhibit low self-concepts, feelings of rejection and helplessness, hostility toward adults, and negative attitudes toward school. They may have poor study habits and social skills, few interests, and low levels of aspiration, motivation, and self-discipline. Underachievement frequently leads adults to considerable anguish and frustration while causing children to experience guilt over disappointing the significant adults in their lives. Patterns of underachievement often are clearly identifiable by the middle elementary school years, and frequently they intensify as the children progress through school. Underachieving boys, whose emotional maturity develops more slowly than that of girls, outnumber girl underachievers by a wide margin.

Underachievement by gifted students is commonly regarded as a psychological problem, a set of learned behaviors. Terman, in his longitudinal study of the gifted (Terman and Oden 1947), was the first to classify and describe gifted underachievers. He identified four major characteristics that differentiated underachievers from high achievers: (1) lack of self-confidence, (2) inability to persevere, (3) lack of integration of goals, and (4) inferiority feelings. Many studies have helped to establish that underachievement by gifted learners is a long-term problem rooted from early childhood in personality, social, and family problems. Unfortunately, numerous gifted underachievers go undetected.

A school atmosphere lacking in intellectual challenge, with an incompetent or insecure teacher and overemphasis on conformity and perfection, can greatly aggravate the problem of underachievement. Special programs that feature stimulating intellectual content and emotional support for the student, including attention to improving poor self-concept and emphasis on motivating achievement, have proved to be successful with gifted underachievers (Gallagher 1975). Among all the strategies that have been reported, however, the two most successful approaches appear to be counseling as well as classroom intervention. Because of the stubborn nature of the underachievement problem, long-term counseling that involves the entire family usually is required.

Social and Cultural Issues

There are no simple prescriptions for resolving the difficulties presented by the social and cultural issues identified above. What is essential is that educators be sensitive to their existence and to the real but often subtle impact they can have on the success of a program. Gifted children must be treated as individuals, each with a unique configuration of psychological and social needs; unless these are taken into account, the likelihood of the child's success and growth in the program is severely diminished.

Organizational Alternatives for Gifted Programs

Discussions of what kind of educational programs to establish for gifted students inevitably lead to some measure of debate over the relative merits of acceleration versus enrichment. Here acceleration is understood to mean progress through the curriculum at a pace more rapid than that of the average student, although the degree of acceleration may vary from a slight advantage, such as completing the year's work earlier than the rest of the class, to the radical telescoping of the standard secondary school mathematics program into one or two years or even less. Enrichment is a more elusive concept that is interpreted differently in various situations, but enrichment programs generally have in common the feature of expanding beyond the content of the ordinary curriculum, an expansion that may range from the introduction of irrelevant topics or "busy work" to the inclusion of significant mathematics studied in great depth and at a higher than average cognitive level.

The danger in any debate over enrichment versus acceleration is the tendency to polarize or dichotomize, as though one must choose either enrichment or acceleration, but not both. In the discussion of content for the mathematically talented that follows in a later section, it becomes clear that programs for the mathematically gifted must have the characteristics of being both significantly enriched and appropriately accelerated. This is equivalent to saying that for the gifted student, both the standard curriculum and the instructional delivery system must be modified.

NCTM's view of the approach to be preferred in programs for the mathematically gifted was set forth in the *Agenda for Action*, which stated that

in general, programs for the gifted student should be based on a sequential program of enrichment through ingenious problem solving opportunities rather than through acceleration alone. (NCTM 1980).

Later, the Council reiterated and expanded this belief in an official position statement that concluded

that vertical acceleration be considered only for a limited number of highly talented and mathematically creative students whose interest and attitudes clearly indicate that they have the ability and perseverance to complete a carefully designed sequential curriculum. For all but this select group, a strong, expanded program emphasizing mathematics enrichment is preferable. (NCTM 1983)

Most recently the Council spoke out on the importance of appropriate programs for the gifted in a 1986 position paper that recommended

that all mathematically talented and gifted students should be enrolled in a program that provides a broad and enriched view of mathematics in a context of higher expectation. Acceleration within such a program is recommended only for those students whose interests, attitudes, and participation clearly reflect the ability to persevere and excel throughout the entire program. (NCTM 1986)

The entire text of this position paper can also be found in the Appendix.

But whatever the nature of the curriculum for the gifted, educators must consider as well the organization of the program through which that curriculum is delivered. Such alternatives can range from modifications within the individual classroom or grade to programs operated on the local school, school district, regional, or state levels. Below are brief descriptions of the most common alternatives; later sections of this publication will elaborate with examples of each.

Classroom-Level Options

For the teacher with one or a few gifted students in an otherwise heterogeneous class, alternatives commonly employed include special projects, research reports, independent study, small-group work, and peer teaching. In some schools gifted students are supplied with an alternative textbook or supplemental learning materials that treat mathematics at a higher level and in greater depth. The practice of allowing bright students to work from the mathematics text for the next grade is a fairly common one, but it has serious limitations, since students in such a situation frequently operate largely on their own and rarely experience the high quality of instruction and teacher guidance that are needed for sound mathematical development. Too often what results is reasonable (or even very good) facility in algorithmic manipulation, but minimal understanding of concepts and principles. Indeed, the problem of minimal supervision by, and interaction with, the teacher is the most serious constraint on any attempt to accommodate gifted students in regular classes.

Local School Options

Greater flexibility is gained when accommodation is made at the school level rather than at the classroom level. One of the most common and easiest to implement options is to enroll pupils in mathematics classes intended for older students. This may mean sending an elementary pupil to the next grade for mathematics or allowing a secondary school student to accelerate progress through the mathematics program. For some students it may also involve going to another school, such as elementary pupils to the middle school, junior high school pupils to senior high school, or high school students to a local college. This is ordinarily accomplished by arranging a schedule that allows for early dismissal or late arrival. Arrangements such as these usually do not involve significant modifications to the standard curriculum, only the opportunity to study "ordinary" mathematics earlier.

More opportunities to tailor the curriculum to the needs of the gifted result when students are grouped by ability. Ability grouping permits teachers to cover content more rapidly, in greater depth, and with greater degrees of abstraction, formalism, and rigor. More nonstandard topics can be introduced; more challenging problems and applications can be presented. Students also have more opportunity to discuss mathematics with peers who can stimulate and challenge their ideas and to pursue topics of individual interest.

School-level provisions for the gifted may place mathematically talented students in special classes only for mathematics or a few subjects; the gifted may be grouped together for academic subjects during part of the day but enrolled in homogeneous classes for the rest of the time; or the school may operate a school-within-a-school program that enrolls gifted students on a full-time basis and that probably operates on its own time schedule quite independent from the rest of the teachers and students.

Magnet Programs

Many school districts operate magnet schools that emphasize a particular academic focus such as mathematics and science, foreign language, or the performing or fine arts. It is not uncommon for school districts of sufficient size to operate several magnet schools with a different emphasis at each site. Attendance is determined by selection criteria that may include demonstrated ability, talent, interest, or potential, or magnet schools may be open to any student upon application; magnets are not restricted by conventional geographical boundaries.

Magnet programs are not necessarily programs for the gifted, although they may attract students with a high degree of talent or interest in the designated field of study. A school with a magnet program in mathematics is likely to offer greater course selection, to have more highly qualified mathematics teachers, and to enjoy more resources for teaching mathematics than other schools in the district, but students in the school probably will not experience a similar quality of educational opportunities in other subjects taught there.

Magnet schools rarely exist outside of large school districts, since small districts lack the critical mass of students and teachers required to sustain a spectrum of magnet programs. Where magnet schools do exist, transportation usually is a complex and costly issue. Unless well established and valued in a school district, magnet schools can find themselves vulnerable to the rise and fall of public support and voter attitudes, especially in times of a shrinking economy.

Special Schools

More comprehensive than magnet programs are certain special schools designed to emphasize a given focus throughout the curriculum for all students. A magnet school may offer an enriched mathematics curriculum for a certain subset of the students in the school; a special school for mathematics will reflect a focus on mathematics throughout the academic program for all students in that school. Such schools usually have more rigorous entrance standards than do magnet schools. Commonly a special school for mathematics will also emphasize science and technology. Two well-known examples are the Bronx High School of Science, operated by the New York City school system, and the North Carolina School of Science and Mathematics, operated on a statewide basis.

Special schools for the gifted also exist. Such schools, many of which are private schools, enroll students with the common characteristic of having superior abilities in one or more areas and the capability to learn more rapidly and in greater depth, but the nature of the students' abilities and interests may be very diverse. A student enrolled in such a school on a full-time basis would expect to encounter advanced learning situations in all subjects, although an individual gifted student is not likely to be equally superior in all areas. Thus, although the content of the curriculum and the pace and method of instruction may be modified for the gifted, the mathematics offerings in this case are likely to be less intense than in a special school for mathematics, since not all the students will be particularly gifted in that area.

Like magnet schools, special schools face limitations of logistics and transportation, extra costs, and the requirement of sufficient numbers of students and teachers. If anything, these problems are more severe for the special schools, and they are complicated by the fact that many taxpayers consider such schools to be "elitist" and therefore unnecessary or even undesirable.

Learning Centers

Learning centers can operate on a small scale within an individual classroom or on a larger plan serving a building, the district, or a region. Small centers offer students an opportunity to interact with printed materials, manipulatives, calculators, computers, audiovisual resources, and other learners and teachers to explore topics of mathematics not otherwise included in the curriculum. Regional centers may offer seminars or short courses that students can attend outside the regular school program, and they have facilities not available in local schools. Computer centers and research laboratories, as well as more conventional curriculum centers, may serve as learning centers to which teachers can take groups of students on a limited basis. Such centers also can support the classroom teacher with resources or personnel to extend the ordinary mathematics program.

Special Classes

Special classes for gifted mathematics students generally operate outside of, and in addition to, the regular curriculum. For elementary pupils these are usually in the form of pull-out programs that remove gifted pupils from their homeroom on a periodic basis for enrichment activities in mathematics that quite likely are not related to the standard curriculum. For junior and senior high school students, the alternative that has gained most popularity in recent years is the fast-paced class.

Fast-paced classes are offered in locations across the nation, some operated by local districts, others by state departments of education, and still others by colleges and universities. Most are modeled on the original Study of Mathematically Precocious Youth introduced at Johns Hopkins University in the 1970s (Stanley, Keating, and Fox 1974). Typically classes meet outside of the school day on Saturdays or in the late afternoon; a common schedule features a two- or three-hour class once a week. Most such classes are held at a central location and draw students from anywhere within commuting distance.

Students enrolled in these classes are frequently sixth through eighth graders who are identified by means of a talent search in which pupils who

3. Mike's house is on the corner of 16th St. and Kansas Ave; his school is on 12th St. and Michigan Ave. Assuming that Mike always goes straight home by the shortest path (but stays on the sidewalks at all times), how many different routes can Mike take from school to home?

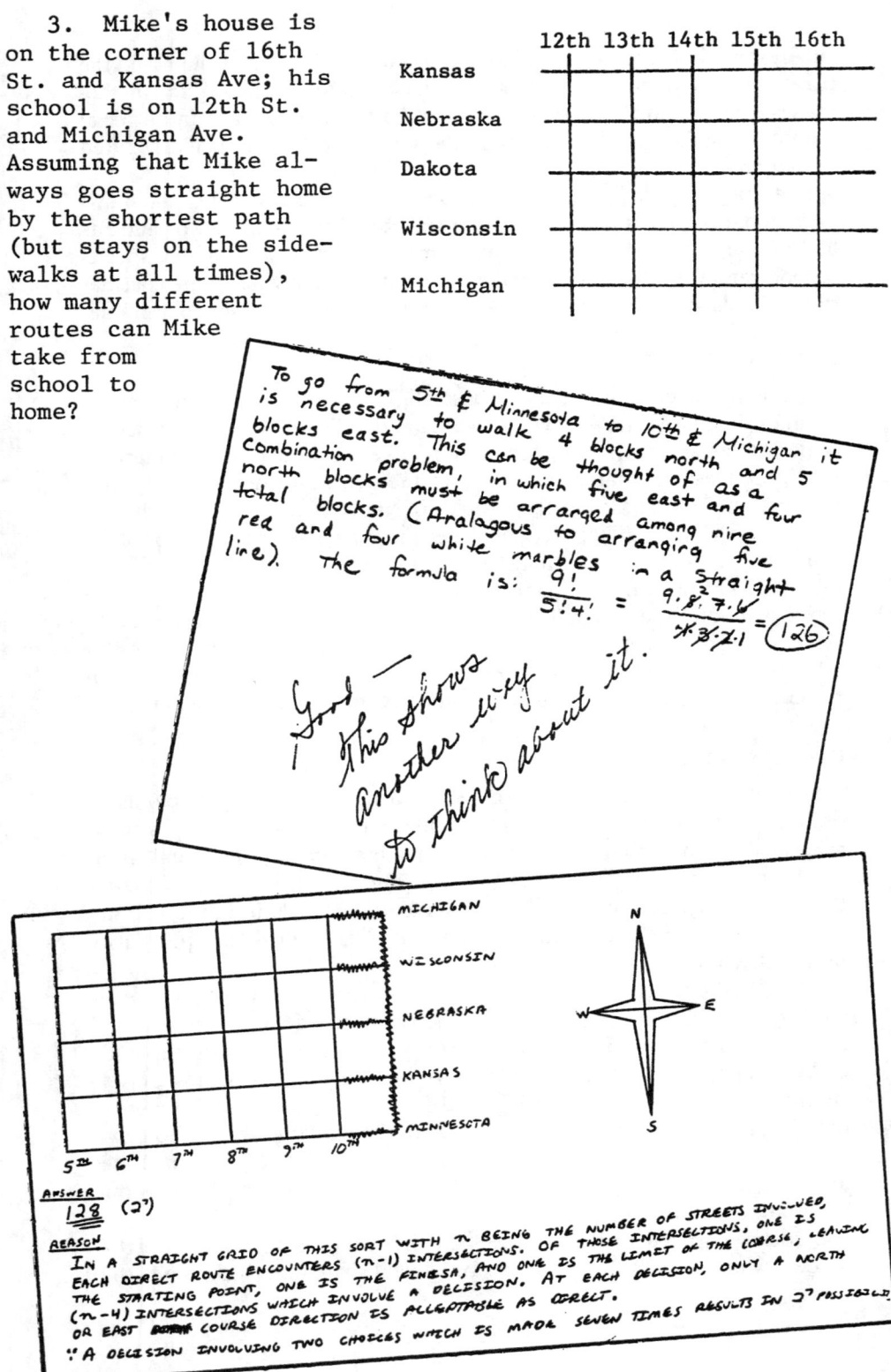

To go from 5th & Minnesota to 10th & Michigan it is necessary to walk 4 blocks north and 5 blocks east. This can be thought of as a combination problem, in which five east and four north blocks must be arranged among nine total blocks. (Analagous to arranging five red and four white marbles in a straight line). The formula is:

$$\frac{9!}{5! \, 4!} = \frac{9 \cdot 8 \cdot 7 \cdot 6}{4 \cdot 3 \cdot 2 \cdot 1} = \boxed{126}$$

Good — This shows another way to think about it.

ANSWER
128 (2⁷)

REASON
IN A STRAIGHT GRID OF THIS SORT WITH n BEING THE NUMBER OF STREETS INVOLVED, EACH DIRECT ROUTE ENCOUNTERS (n-1) INTERSECTIONS. OF THOSE INTERSECTIONS, ONE IS THE STARTING POINT, ONE IS THE FINISH, AND ONE IS THE LIMIT OF THE COURSE, LEAVING (n-4) INTERSECTIONS WHICH INVOLVE A DECISION. AT EACH DECISION, ONLY A NORTH OR EAST COURSE DIRECTION IS ACCEPTABLE AS CORRECT.
∴ A DECISION INVOLVING TWO CHOICES WHICH IS MADE SEVEN TIMES RESULTS IN 2⁷ POSSIBILD

have scored very high on in-grade tests of mathematical aptitude or achievement are tested with a more advanced instrument, probably one designed for entering college students. The program, an example of radical acceleration, is fast paced and difficult. A popular curriculum model offers two years of high school algebra during the first year of special classes; pupils can continue for a second year of geometry and trigonometry leading to subsequent courses in which one or two years of calculus are taught to students while they are still in high school.

Some concerns associated with such special classes affect the participants directly. The record of fast-paced classes in the past shows that they have not been equally successful with boys and girls. Usually the number of girls who qualify for these classes is far below their proportion in the talent searches from which they are selected, and the attrition rate has been especially high for those girls who do begin the program. Also, classes like these require large amounts of homework between sessions, and pupils do not see the teachers on a daily basis. Thus, to be successful, participants must be highly motivated and capable of sustained independent work at a high-intensity level. Tests of mathematical knowledge or ability do not measure such qualities. However, available evidence supports the conclusion that pupils in fast-paced classes do master the content of the standard secondary mathematics curriculum, though it is not clear that they necessarily develop good problem-solving skills.

Some programs have adopted the special class format but have rejected the notion of radical acceleration and employed instead an enrichment approach to secondary and college level mathematics. In these classes students are likely to find an integrated curriculum built around text materials such as *Unified Modern Mathematics* (Fehr et al. 1968–1972) or *Elements of Mathematics* (Martin 1970–1983). Arrangements are made that allow participants to receive high school credit for some of their work and college credit for the rest.

Whether accelerated or enriched, special class programs must contend with a number of unique problems in addition to the obvious financial and logistical ones. When they operate outside the jurisdiction of the local school, public relations and communication with the home schools can be major obstacles. It is not uncommon to encounter resentment from the local schools and teachers who perceive the special classes as "robbing the schools of the best students." School personnel may object to the administrative inconvenience of making provisions for a pupil, perhaps as young as ninth grade, who has already completed the entire secondary mathematics program; and it is a serious disservice to accelerate a pupil through the curriculum only to leave him or her faced with more required years of schooling and no mathematics to take. In a few cases, schools have been reluctant or even refused to grant high school credit for courses taken in the special programs, and it is not uncommon for home schools to require pupils from such programs to take the regular junior high school courses even though they have already completed much more advanced work, a practice which is educationally indefensible.

Summer Schools

Summer alternatives for the gifted run the gamut from ordinary academic offerings that allow students to earn credit sooner and possibly gain early admission to high school or college to programs that provide opportunities to study content not otherwise available in the standard school curriculum. Summer programs may or may not be specifically designed for the gifted. They may be either day school or residential, the latter being more likely to be selective and focused on the gifted. Several states sponsor summer academies in the form of residential programs held on college campuses that bring together selected gifted students and outstanding teachers in an environment that includes college-level instruction, exposure to new branches of mathematics, opportunities to interact with similarly gifted peers, and access to special programs, guest lecturers, and the research facilities of the host university.

Mentor Programs

Alternatives for the gifted need not be restricted to formal instruction in classes or schools. The mentor alternative is an example of such an option. In a mentor program, individual gifted students are paired with adult professionals in the community who can serve as guides, motivators, tutors, and role models. The student works with the mentor on a prearranged, usually regular, schedule either during released time or outside of the school day. Sometimes the mentorship extends to a work-study opportunity during vacations and summers. Mentors are almost always volunteers, so costs are relatively low compared to other programs for the gifted, but arranging such relationships for more than a very few students can be time consuming and even prohibitive. Also, the learning outcomes from mentor programs are virtually impossible to predict, since there is generally no formal curriculum or evaluation involved.

Clubs and Competitions

Although not a program with a formal curriculum, mathematics clubs and teams can provide other opportunities for talented students to engage in mathematical experiences outside of school. Clubs may sponsor speakers, field trips, fairs, and other special projects of interest to the students. Preparing for mathematics competitions, whether entered individually or as a member of a team, is motivating and challenging to many talented students who, in the course of their preparation, also may learn both new content and more sophisticated mathematical problem solving processes.

Considering Program Alternatives

Because gifted students are as diverse a population as any group of pupils, they will vary widely in interest, ability, and learning style. No one alternative is best for every child, just as not all options are workable in every situation. Thus programming for the gifted must include, along with a clear definition of the desired population and a careful delineation of selection procedures, a thoughtful weighing of the programming alternatives to determine which ones are best suited for each situation.

Mathematical Content for the Gifted

The literature describing the content of school programs for the gifted is extensive, but unfortunately it offers very little with regard to specific academic content, especially at the secondary level. Experts have produced volumes debating pacing, learner independence, creativity, and underlying philosophies of gifted and talented programs, and they have described programmatic approaches based on one or more of the following principles:

- An elementary-secondary-college curriculum continuum where students accelerate their studies beyond those of their age cohort
- A typological continuum derived from identifiable kinds of giftedness
- A topical extension or excursion where specific topics are added to the standard curriculum
- A continuum of teaching strategies based on specific ideas about the way the mind is structured, or learning taxonomies, or on strategies for encouraging creativity, such as brainstorming
- A continuum based on what can be accomplished, generally in a setting that includes local support, high energy by a few key people, and limited resources

Some have described generic content modifications for the gifted, almost always dealing with elementary school programs; *Teaching the Gifted Child* (Gallagher 1985) is a good example. Exemplary programs for talented students are publicized through a variety of organizations, including NCTM. A recent example is *The Secondary School Mathematics Curriculum* (Hirsch 1985). Studies of the achievement of individuals, such as *Developing Talent in Young People* (Bloom 1985), *Educating Able Learners* (Cox, Daniel, and Boston 1985), and *Smart Girls, Gifted Women* (Kerr 1985), identify significant aspects of the education of able learners. Practitioners writing in books like *Respecting the Pupil: Essays on Teaching Able Students* (Cole and Cornell 1981) share the benefits of their experiences. *Teaching Models in Education of the Gifted* (Maker 1982) presents a useful source of information about the philosophical theories of Bloom, Bruner, Guilford, Kohlberg, Parnes, Renzulli, Taba, Taylor, and Treffinger.

Specific classroom activities can be found in journals, the most notable for mathematics instruction being the April 1983 *Mathematics Teacher* and the February 1981 *Arithmetic Teacher*. "Survival guides" tell teachers, parents, and students how to deal with giftedness; and the ERIC Clearinghouse on Handicapped and Gifted Children publishes occasional digests on topics such as definition and identification, delivery of services, instructional alternatives, parents' roles, and teacher training. Disappointingly, though, many high-visibility reports such as *A Nation at Risk* (1983), *Action for Excellence* (1983), and *High School* (Boyer 1983) failed to make plain the relationship between the need for academic excellence in our schools and the need for special provision for those students having the greatest potential. Unfortunately too, although the material mentioned above is interesting, it may have little specific information about mathematics content to offer to teachers who hope to improve the instruction they provide for their gifted students. In a concise phrase: many pages—little help.

What, then, are the content issues facing those interested in providing a quality mathematics program for their gifted students? We can identify the following:

1. Misdirection and inappropriate expectations by generalists who provide the bulk of the literature
2. Lack of preparation in mathematics by many teachers
3. Lack of materials including the textbooks, teachers' manuals, and supplementary material that are appropriate for both the ability and age of the students (e.g., junior high school students can do abstract algebra but not from a college textbook)
4. Lack of articulation between grade levels and a consequent lack of continuity
5. Lack of guidance in the selection of content and expectations for students

Nevertheless, despite such constraints and a paucity of helpful documentation, when we consider the advice of experts, look at program prototypes, talk to practitioners, examine the characteristics of able learners, listen to students, and think carefully about what we hope to achieve, we discover that we can make some useful generalizations and can suggest the design of reasonable, workable, and effective programs. But before we discuss what the content of programs for the mathematically gifted should be, let us enumerate those provisions that are, by themselves, insufficient:

- Friday "fun with math" time
- Field trips
- Brainstorming or creativity sessions
- Independent projects
- Independent study
- Serving as helpers for other students
- Using computers

Let us be clear: there is nothing wrong with any of the activities above, but they simply do not constitute a program. In fact, such activities tend to cloud our focus by creating the illusion that the needs of gifted children are being met. For example, "Friday mathematics" fosters a dichotomy between important mathematical ideas and what students come to see as the "real content" of the curriculum on which they are tested and graded. Similarly, although brainstorming sessions can be exciting, they can also have the dangerous side effect of fostering a superficial, dilettantish quality that allows serious problem solving to be cast aside. Furthermore, when properly implemented, these activities are appropriate for all students, and it is indefensible to suggest that they are appropriate only for the gifted.

It is equally unreasonable for bright students to do long "busy work" assignments or to repeat content they have already learned. Instead, programs for gifted and talented students should allow participants to use their strengths to maximum advantage. They should also allow for intellectual and personal growth in a manner that is unique to the capabilities of bright children and that would be inappropriate for their less able classmates.

Curriculum is the medium through which learning occurs, and the special abilities of the gifted and talented require programs that provide opportunities to develop abstract thinking, to sharpen higher cognitive processing, to practice creative problem posing and solving, and to enlarge individual methods and styles of inquiry. A solid program requires long-term planning beyond the scope of one teacher in a single classroom or even of a group of teachers at a single grade level. Students will benefit most from a serious, long-range commitment by professional staff, parents, and the students themselves.

Challenging programs for the gifted should help students to deal with the frustrations and rewards of answering questions that they once viewed as unanswerable. Many very bright youngsters have no sense of how to address problems that are new to them, even though they do their standard schoolwork exceptionally well. Many, if presented with a problem that they cannot answer immediately, respond with withdrawal. When the going gets rough, as it should for gifted students, they must be given appropriate support and direction. They deserve high-quality instruction in the material they are learning as well as guidance and firm encouragement for good work habits and study skills. And programs for the gifted should excite and motivate students to independent investigation. Gifted students are not necessarily self-starters.

Bright students must learn to be articulate and precise about their own ideas and encouraged to explore unselfishly the ideas of others. Their stamina must be stretched beyond playfulness, and they should be routinely expected to write well, think well, and speak well—to use both good mathematics and good English. When necessary, their errors should be corrected and their work revised until it meets a reasonable yet rigorous standard.

Students should also become epistemologists to gain not only a good sense of their own thinking but of the individual differences that will surely exist among the students in any such program. The acts of answering the same question by using different techniques and considering a problem that has more than one answer are almost always viewed skeptically by youngsters who have been schooled in an environment where answers are valued above all else and where problems are strictly categorized to allow for only one correct solution process and result. Variety and alternatives are seldom emphasized, and sometimes not even tolerated, in typical classrooms.

The mathematical content of programs for gifted and talented students should be enriched in three ways: It should include standard topics in greater depth (e.g., error analysis when studying decimals); additional nonstandard content (e.g., non-Euclidean geometry, graph theory); and early study of advanced content (e.g., indirect proof in middle school). Further, it should reflect contemporary mathematical concepts and curricula (e.g., doing mathematics with calculators and computers) and stress the importance of process. Gifted students need to see dimensions of the richness of mathematics that are generally too sophisticated and subtle to be understood and appreciated by their less able classmates. Their mathematics should be organized and written in the style of contemporary mathematicians and those who use it. It should have a high degree of abstraction and notation, and go beyond mere symbol manipulation to underlying principles and proof.

The practice of placing bright students of grade x into a class of average students of grade $[x + a]$ (and bright students of grade $[x + a]$ into grade $[x + 2a]$) is an inadequate response, since gifted students require instruction that is qualitatively different from that appropriate for average classes. By the same token, junior or senior high school youngsters placed into standard classes of college students find themselves no more than oddities who are, once again, not receiving instruction specifically geared to their abilities. Bright students are just that, and forcing them to be diminutive adults or miniature mathematicians robs them of many important aspects of their childhood.

Talented students want to be taken seriously; they want someone to carefully scrutinize their thinking. They desire verification when they are correct and an explanation when they are wrong. They are willing to make mistakes, but they are at the same time generally unwillingly to take chances with their own ideas unless someone will listen carefully to their sometimes long, involved arguments. Bright students seek opportunities for knowledgeable dialogue, and it should be provided.

Last, and probably most important, any program for gifted and talented students must have academic integrity. Students should be held accountable for knowing more because they are in a special program, and the content they are expected to know must have more substance than piecemeal topics of mere curiosity. Mathematics by its very nature builds on itself, and its powerful ideas are embedded in many of the basic notions that are routinely taught in elementary and secondary school. The curriculum for the mathematically gifted must communicate significant mathematics in an interesting and effective way so that the students not only learn more mathematics but, in the process, develop their thinking skills, thus becoming productive and capable of continued intellectual and personal growth.

The Instructional Environment for the Gifted

Since the publication of the *Agenda for Action* in 1980, problem solving has gained widespread acceptance as the most important goal for mathematics instruction:

The development of problem solving ability should direct the efforts of mathematics educators through the next decade. Performance in problem solving will measure the effectiveness of our personal and national possession of mathematical competence. (NCTM 1980, p. 2)

Since problem solving emphasizes higher-order thinking skills, it is the ideal vehicle for enrichment, challenge, and sophistication in the mathematics classroom. Problem solving should be a fundamental basic skill to be developed by all students, but it is especially critical in the mathematical development of the talented.

Gifted students should learn in an environment where they are presented with significant problems and given enough time to work on them without interruption. In this environment, the first priority should be for quality and sophistication of instruction, not lonely learning. There must be an appropriate level of formalism in the problems, projects, and even instructional games, and this formalism should be developmental in its degree of sophistication. Gifted students sometimes require considerable intellectual prodding.

There is a popular tendency to assume that gifted students can learn easily on their own with the teacher merely acting as organizer and evaluator of their learning. On the contrary, teachers have a responsibility to the gifted no less than to other students to continue to interact with them at appropriate times during the solution of a problem or the development of a project, to pique their curiosity and challenge them to defend, clarify, and generalize their thinking.

Problems and topics for investigation that involve several areas of mathematics should be chosen so that solutions can be generalized and extended to other concepts or topics. Schlesinger (1983) described an example of such a problem solving lesson in which a problem to determine a given social security number involved divisibility tests, required extensive logical thinking, and led to computer applications. Polya (1981, vol. 1, pp. 68–73) showed how a problem that begins as a counting task can lead to Pascal's triangle and further generalizations that apply to other path counting problems. These are examples that illustrate the manner in which students should routinely be encouraged to expand problems and generalize solutions. They also should be encouraged to develop problems for themselves and others.

Problems that encourage divergent thinking and creative, clever solutions are important. Students should be encouraged to strive for brevity and elegance in their solutions and to find more than one approach to a given problem. Educators must develop an alertness for the original and unique approaches of the gifted and capitalize on them to stimulate further thinking.

Girls, even gifted girls, tend to be less successful in problem solving than boys. Tobias (1978) credits this tendency to the fact that problem solving involves self-confidence and risk taking, and she notes that girls are socialized to be less confident, poorer risk takers than boys. One strategy to combat this situation is to use cooperative groups in problem solving. Many students speak more readily and contribute ideas more freely in small groups than in front of an entire class. Ridge and Renzulli (1981) also note the affective gains to be realized by the gifted during group problem-solving experiences including the development of self-concepts and learning to listen to and appreciate the ideas of others. All students must be helped to realize that mathematics learning demands the freedom to make mistakes and to learn from them. Gifted students are not good at accepting their mistakes; they require a learning environment that supports the freedom to learn from their errors.

Concrete instructional materials are widely accepted for teaching slow learners, but many educators overlook their effectiveness for providing additional insights and helping the gifted as well to realize that mathematics does not take place solely on an abstract level. Advanced topics from geometry to trigonometry, conic sections to functions and more, require demonstration on a concrete level. Mathematical games such as chess, Mastermind, Nim, and others can also be of value in the mathematics classroom. Besides adding variety to the classroom routine, they provide practice with logic and visual skills and develop group interaction and learning.

Students also must be given valid and challenging applications of mathematics, since "the whole dimension of mathematics developed from, and applied to, practical considerations is one to which gifted students need considerable exposure if they are to grasp the concept of mathematical modeling and simulation—the very heart of the mathematics of today's and tomorrow's business and industrial procedures" (Ridge and Renzulli 1981). Many excellent application problems can be chosen that stress the fusion of mathematics with a wide variety of other disciplines (see, for example, Bushaw et al. 1980; Tanur et al. 1978). Applications, both "real world" and fanciful, provide an excellent way to merge the learning of mathematics with students' other academic interests and to allow them some choice in the material they study.

A particularly valuable source of application materials is the Consortium for Mathematics and its Applications (COMAP), a nonprofit corporation that develops and sells materials in print, microcomputer, and video form for use in science and mathematics classes in secondary schools and teacher education programs. It also publishes *Consortium*, a newsletter that features application-based problems, articles, and lessons that are ideal for gifted high school students.

There also has been a developing trend to stress the importance of situational lessons that explore and solve significant real-world problems. An example of a situational lesson would have students consider a specified parcel of land and design a parking lot that can most efficiently and economically accommodate cars of various sizes, buses, recreational vehicles, and trucks in proportion to their representation in the universe of motor vehicles. The lesson involves a great deal of mathematics, including geometry, statistics, modeling, computations, and decision making. If the problem addresses a personal concern of the students, such as providing adequate parking near the school, it takes on added significance for them. Other problems can be highly imaginative, such as designing school or recreational facilities for a twenty-first-century space station.

To support learning environments such as those suggested above will require deviation from the traditional teacher-centered, teacher-directed classroom structure. The cooperative learning environments found in open classrooms and facilities for independent study will be essential components of gifted education programs. Clark (1983) described successful classrooms for the gifted as being structured around the following:

- Cooperation between students, teachers and parents
- A flexible, integrated curriculum
- An environment more like a laboratory or a workshop that is rich in materials and emphasizes experimentation and involvement
- A minimum of total-group lessons in favor of an emphasis on small-group lessons and cooperative learning
- Evaluation of student growth based on assessments, contracts, and self-evaluations
- An atmosphere of trust, acceptance, and respect

Clark also noted that recent research studies have discovered that because gifted students develop a locus of control at a younger age, they experience improved academic achievement and enhanced self-concepts when they are given choices and some control over their learning environment. Also, gifted students usually become involved in learning and problem solving for the pure pleasure of it; therefore, programs should be flexible enough to allow them to make decisions and real choices about their learning experiences.

Regardless of the classroom setup, it must be organized and well planned with a systematic and businesslike approach, but it must also be a place with a warm, safe, and permissive environment. Whatever formal instruction is presented in the classroom should be based on the lecture method as little as possible. Instead, the discovery or guided-discovery approach should predominate. (Polya [1981] presents an extensive discussion of discovery.) Do not expect, however, that students accustomed to being told what to do and how to do it will necessarily embrace discovery lessons with enthusiasm. Their impatient "just tell me how to do it" requires a special determination on the part of the teacher as well as a measure of both patience and ingenuity, especially when students are first exposed to this kind of learning environment.

It cannot be assumed either that gifted students are automatically proficient in the basic skills of mathematics. Yet there is a tendency in some cases to push the mathematically gifted into accelerated classes and possibly to omit some critical learning from the curriculum. Teachers must be careful to provide periodic review of fundamental concepts and skills, to monitor curriculum and achievement, and if necessary, to review important topics that may have been omitted previously. Since many gifted students are bored by, or resistant to, repetitive drill and routine learning, it is important to provide a delicate balance between necessary remediation and more creative activities. Many commercially available motivational drill and practice materials can be helpful in bridging this gap.

It also must be remembered that learning mathematics is not based only on symbolic and logical components; it is also highly verbal in nature. Students must develop an awareness of their own mathematical thinking and must learn to communicate their thought processes verbally and in writing. They must learn to submit their ideas to public discussion and scrutiny. Students can learn considerable mathematics by explaining their ideas and defending their thought processes to others. Furthermore, the learning of mathematics also involves skills such as visualization and spatial perception, which are too often deemphasized in traditional mathematics curricula. For two especially fascinating references, see Martin Gardner's *The Ambidextrous Universe* (1979) or *Experiences in Visual Thinking* (McKim 1980).

In summary, the learning environment for the gifted was aptly described by Kaplan (1974, p. 8) as follows:

A program for the gifted and talented provides multidimensional and appropriate learning experiences and environments which incorporate the academic, psychological, and social needs of these students. The implementation of administrative procedures and instructional strategies which afford intellectual acquisition, thinking practice, and self-understanding characterize a program for the gifted and talented. A program assures each student of alternatives which teach, challenge, and expand his knowledge while simultaneously stressing the development of an independent learner who can continuously question, apply, and generate information.

Evaluating Students and Programs

Evaluating educational programs for gifted students is a complex and difficult undertaking, due in large measure to the uniqueness of each program's design and the multifaceted nature of giftedness.

The purpose of evaluating any educational program is to find out what is working and what needs to be improved. Evaluation should be an integral part of the design of the program and should be closely tied to the program's objectives and rationale. The ultimate goal and outcome of the evaluation should be to produce growth in the students and improvement in the quality of the program. Evaluation should never be an end in itself. It should be conducted systematically over the course of the program's duration, and a formal summative evaluation report should be produced. For ongoing programs this is usually done on an annual cycle.

The basic question underlying the evaluation process is "Who needs to know what?" Teachers, parents, and pupils must have an assessment of pupil progress. Teachers and school officials require assessment of the structure and general atmosphere of the classroom. Evaluation must consider the needs of students, teachers, parents, administrators, school board members, and sometimes funding agencies, not all of whom may agree on what constitutes a proper program for the gifted. Each group should be given input into what they think needs to be evaluated.

It is also necessary to consider the amount of time, money, and other resources that will be available for the evaluation. Appropriate instruments such as rating scales, tests, or questionnaires must be purchased or designed in order to obtain the required data, which must then be analyzed by the use of appropriate logical and statistical procedures. Finally, results must be communicated to all concerned parties, and appropriate program modifications based on the results of the evaluation must be implemented. The evaluation system itself must be periodically monitored in order to make recommendations for improvements and refinements in the system.

If the gifted program is part of the curriculum of a large school system, evaluation may be conducted by an independent evaluator or team from the district office. However, a separate, school-based evaluation is also desirable to provide additional feedback to parents and pupils. If evaluation is approached in the spirit of a self-study, it can provide realistic and meaningful information. Whether or not the district has an office of evaluation to conduct the assessment, it is also advisable to employ an external evaluator in order to assure objectivity and add credibility to the findings of internal reviewers.

A companion issue to evaluation is the matter of good public relations and communication about the achievements of the gifted students during the course of the year. Parents, teachers, administrators, school board members, and the public should learn of the awards, publications, creative works, performances, or other accomplishments of the students as they occur.

One common measure of student progress is the traditional system of letter grades. This may appear to be a convenient way of evaluating student learning, but research has shown that letter grades are low in both validity

1. ABCD is a square with M the midpoint of DC. Find the ratios of the areas of the regions P, Q, R, and S.

I simply wrote formulas for what I did know, and then solved them. Let $s, r, p,$ and q = four respective areas. l = length of a side

Since $\overline{AC}$ is a diagonal, ABCD is divided into two $45°-45°-90°$ triangles. Therefore

① $q + r = \frac{1}{2} l^2$

② $p + s = \frac{1}{2} l^2$

Since M is the midpoint of $\overline{DC}$, $MC = \frac{1}{2}l$. Therefore

③ $r + s = \frac{1}{2}(\frac{1}{2}l)(l) = \frac{1}{4}l^2$

In triangles $\triangle AXB$ and $\triangle CXM$, $m\angle A = m\angle C$; $m\angle X$ on $\triangle AXB = m\angle X$ on $\triangle CXM$ (vertical angles). Therefore, they are similar, and corresponding parts are in proportion. Since $\overline{AB} = 2\overline{MC}$, the area of $\triangle AXB$ is 2^2, or 4, times as large as $\triangle CXM$. Therefore

④ $q = 4s$

sub ④ in ① $r + 4s = \frac{1}{2}l^2$
③ $r + s = \frac{1}{4}l^2$
subtract $3s = \frac{1}{4}l^2$
⑤ $s = \frac{1}{12}l^2$

sub ⑤ in ② $p + \frac{1}{12}l^2 = \frac{1}{2}l^2$

$p = \frac{5}{12}l^2$

$m_w = -y/x = -1$
$m_v = y/x = 2$

Equation w: $y - 0 = -1(x-1)$ or $y = 1-x$
Equation v: $y - 1 = 2(x-1)$ or $y = 2x-1$

Finding (x,y)

$y = 2x - 1$
$y = -x + 1$
$0 = 3x - 2$ therefore $x = \frac{2}{3}, y = \frac{1}{3}$

$S = \frac{1}{2}(bh) = \frac{1}{2}(\frac{1}{2} \cdot \frac{1}{3})$
$= \frac{1}{12}$

$R = \frac{1}{2}(b \cdot h) = \frac{1}{2}(1 \cdot \frac{1}{3}) = \frac{1}{6} = \frac{2}{12}$

$Q = \frac{1}{2}(b \cdot h) = \frac{1}{2}(1 \cdot \frac{2}{3}) = \frac{1}{3} = \frac{4}{12}$

$P = \frac{1}{2}(b_1 h_1) + \frac{1}{2}(b_2 h_2) = \frac{1}{2}(1 \cdot \frac{2}{3}) + \frac{1}{2}(\frac{1}{2} \cdot \frac{1}{3}) = \frac{1}{3} + \frac{1}{12} = \frac{5}{12}$

The ratio of the areas of regions P, Q, R, and S are 5 to 4 to 2 to 1 respectively.

Assigning points A, B, C, and D points in a rectangular coordinate system, allows you to easily find the point of intersection of $\overline{AC}$ and $\overline{MB}$. From there, it is easy to calculate the areas of the various triangles.

and reliability, that they do not promote learning, and that they are poor indicators of what really has been learned. Even worse, "Under the threat of grades, bright students balk at venturing into the unknown or trying any area in which they are not sure they will succeed" (Clark 1983). Bright students have been known to avoid honors classes that are graded more rigorously than regular classes because they are afraid their grade point averages may be negatively affected. An increasingly popular solution is a system of weighted grades for honors classes.

Evaluation without grades facilitates the learning process (Clark 1983, p. 320). Ideally, evaluation should be a continuous process that makes the students aware of their strengths and weaknesses, interests and abilities, in a positive and nonthreatening way. What are needed are alternatives for enabling students to see whether they have met their own goals. Alternatives to traditional letter grades include teacher observations, conferences, self-evaluation, and self-diagnosis, possibly by means of a contract, checklist, or essay.

Student progress in gifted programs is often measured by standardized, norm-referenced achievement tests. However, such tests are almost never appropriate for the gifted, since gifted students usually score at or very near the test's ceiling and therefore show misleadingly small gains that do not reflect their true progress. Furthermore, standardized tests are usually not designed to discriminate among students who are all performing at nearly the same level (Gilberg 1983). In addition, Gallagher (1985) points out that standardized tests are weakest in the areas where gifted programs place the greatest emphasis: the development of higher-order thought processes.

Gifted students often think they have failed if standardized tests do not indicate that they are equally strong in all areas. They need to be made aware that they have individual strengths and weaknesses and are not expected to excel in everything. Further, the gifted are not immune to test anxiety and, in fact, may experience heightened levels of anxiety because they perceive themselves as having more to lose by poor performance. Finally, it is important that the test focus on collecting only useful and relevant data. Too often gifted students are subjected to inappropriate routine standardized tests that provide little useful information simply because it has been traditional to test all students in such a way at a given age or time.

Other viable measures of student growth in gifted programs include specially constructed tests and rating scales, student interviews, teacher observations, parent and student questionnaires, student self-reports, and group visitations. Renzulli (1975) has provided valuable advice and information on evaluation design. It is important to remember that since gifted students are so perceptive, they are extremely important sources of information about whether a program is meeting their needs, and they can provide valuable suggestions for improvement. Given a chance, they will tell (Clark 1983).

A major problem in evaluating the progress of the mathematically gifted is that programs should place emphasis on creativity and higher-level thinking skills, even though the development of valid and reliable instruments to measure such outcomes is still in its infancy. In the absence of formal tests, one approach is to have students keep problem-solving notebooks and to submit these records of their efforts, both exploratory and refined, to be evaluated and recorded over a period of weeks or months. Together students and teachers can discuss and assess progress after studying these documents.

A second problem in evaluating students in gifted programs arises because of the greater degree of individualization often found in such programs. Obviously, when individual students are working toward individualized objectives, group tests will be an ineffective measure of progress. This fact further argues for the importance of alternative approaches as discussed above.

In reviewing the literature on the evaluation of the gifted, Rogers (1986) summarized the following major problems encountered in attempts to conduct comparative assessments between programs for the gifted and other programs:

1. It is not possible to demonstrate the effectiveness of a gifted program by showing that students in the special group score several grade levels above their chronological peers on achievement tests, since gifted children in regular programs also perform extremely well on achievement tests and there is every reason to believe that they would perform above their age level whatever the program.
2. It is not possible to prove the effectiveness of a program for the gifted by giving achievement tests before and after the program because, even if accelerated educational growth is observed, there is no proof that the students might not have done just as well in a regular program.
3. One cannot demonstrate the effectiveness of a program by obtaining the opinions of people connected with it (teachers, parents, or students) unless those opinions are supported by objective measures of some sort.
4. The benefits of a program for the gifted will not be demonstrated by comparing the students in the program to others in their grade level, because any achievement gained by students in the gifted program may be due not to the program itself but to the large differences in the two groups to begin with.
5. It is not possible to demonstrate the benefits of the special program even when students in the gifted program are matched for IQ with students in a control group, since many other important factors such as motivation also are important determiners of progress.
6. A program for the gifted cannot be evaluated adequately if the measuring instruments are not appropriate for the unique nature of the program.

Because the learning environment in a program for the gifted has about it many unique characteristics, the evaluation with the most to offer will usually be formative evaluation aimed at collecting data to monitor student progress and to guide program improvements. What are important in such evaluation efforts are (1) a clearly articulated statement of the philosophy and goals of the program that is consistent with sound educational practice and with the characteristics and needs of the students to be served; (2) recognition that giftedness is not a unidimensional trait and that there is no one single way to assess progress or achievement; (3) a psychologically safe and supportive environment for assessing student performance and for communicating to students and parents both the results of that assessment and a plan for further development and growth; (4) opportunities for all parties affected by the evaluation to have appropriate input into the process; and (5) a carefully designed and implemented ongoing system for collecting comprehensive data about both student achievement and program effectiveness. After that, the crucial question becomes what you do with the information you have.

Teachers for the Gifted

"Few educational decisions have as much influence on the gifted program as teacher selection" (Clark 1983, p. 365). With this in mind, it is imperative that we examine the characteristics of successful teachers of the gifted.

The consensus of gifted students, researchers, and educators who work with gifted pupils yields a list of traits valued in all teachers, but especially so in teachers of the talented:

The ideal teacher of the gifted mathematics student—

- is emotionally healthy, a real and authentic person who accepts self but is sensitive to others and respects and trusts them; is honest and sincere, flexible and resourceful;
- is energetic and vital;
- has experience and maturity;
- has a strong background in mathematics and a degree of professional involvement;
- has cultural and intellectual interests both inside and outside the field of mathematics;
- demonstrates enthusiasm for mathematics and for teaching; is able to make the subject matter come alive; is an effective communicator who likes to share ideas;
- is student centered and displays a personal interest in students and a belief in their abilities; is a leader and motivator; can listen to students and learn from what they say and do;
- possesses a strong sense of humor and the ability to make learning fun;

- understands the social, emotional, and educational needs of the gifted; has a preference for teaching bright students;
- understands the thinking and learning styles of the gifted, especially when these are different from the teacher's own; accepts new and different ideas and tolerates ambiguity;
- is enthusiastic about seeking knowledge and new ideas; views teaching as a way to enhance one's own intellectual development;
- is sufficiently confident of his or her own abilities not to be threatened by gifted students; resists the temptation to compete intellectually with students, but instead fosters the goal of seeking new knowledge together; values change, growth, and self-actualization for self and for others.

Of course, it is unrealistic to expect that all the characteristics above will be found in any one person, but the list presents ideals for which to strive. Perhaps what is most important in the final analysis is that the teacher can work successfully with the gifted, help them learn according to their high potential, and motivate them to feel good about themselves and their learning.

Although the listed characteristics are meant to apply to all teachers, there are special comments to be made about elementary teachers. Gallagher (1985) contends that many elementary school teachers of mathematics are math avoiders who may suffer from math anxiety. Since most teacher preparation programs require a minimum of mathematics for elementary school teachers, there are relatively few elementary teachers with strength in mathematics. An increase in the mathematics requirements for elementary teachers will not necessarily relieve the problems of math anxiety and avoidance. Teachers also need to continuously improve their mathematics knowledge and teaching skills through in-service activities and additional course work.

Other benefits can be gained from mathematics specialists in local schools—that is, teachers with strong preparation in mathematics who have as their primary assignments to teach mathematics. Team-teaching arrangements that include at least one team member with strong mathematics preparation can also be effective. And there is need for greater communication and interaction among elementary and secondary mathematics teachers for purposes of mutually sharing both content and teaching techniques. By whatever means are available to a school or district, every effort must be made to assure that the mathematics teachers of young gifted students have strong backgrounds not just in arithmetic but in the entire mathematics curriculum, and that they are eager and enthusiastic about teaching it.

There is a movement in some places to grant teaching certification in gifted education per se. Mathematics educators must be extremely cautious of this practice, since the likelihood is high that holders of such certificates will not possess the strong content preparation in mathematics that is imperative in working with the mathematically talented.

George Polya, mathematician and educator, speaking specifically of mathematics teachers, lists "Be interested in your subject" and "Know your subject" as the first two of his ten commandments for teachers. He adds:

> Both interest in, and knowledge of, the subject matter are necessary for the teacher. I put interest first because with genuine interest you have a good chance to acquire the necessary knowledge, whereas some knowledge coupled with lack of interest can easily make you an exceptionally bad teacher. (1981, vol. 2, pp. 116–17).

Persons responsible for gifted education programs must be realistic about the personnel requirements of such programs. As Rogers (1986, p. 15) warned:

> One generalization that can be drawn from the literature in gifted education is that special programming for gifted children requires additional personnel and services. We should disabuse ourselves of the notion that already overworked teachers and administrators can absorb one more special program into the already overcrowded and understaffed general program. It is wishful thinking to suppose that hard working teachers without sufficient content knowledge, without special knowledge of gifted children, without time for planning programs, and with limited assistance from supervisory personnel will be able to alter, in any meaningful degree, the educational situation for gifted children.

Program Operation

Establishing or renovating programs for gifted and talented students can be exciting and rewarding, but you can also expect to be frustrated, disillusioned, and frequently exhausted. Careful planning and effective communication can minimize some obstacles and help create conditions for successful implementation. So where should you begin?

You are no doubt filled with wonderful ideas, but you will need to turn your aspirations into reality. Begin by developing both long and short term plans. What do you expect your program to be in five years? Ten years? What do you expect to be doing next year? Start with a program small enough to manage with success, a vision of the future, and a time line. You can later expand from there. What you need most in the beginning are a good plan and support. Don't try to import someone else's program in toto, but don't get trapped into reinventing the wheel, either. In whatever you do, your own common sense and judgment are important.

A good first step in generating support is to assess your school's existing curriculum and staff. Even if you do not have a formal program for the gifted, there surely are personnel who are working effectively with exceptional students. They can be your first allies, but it is important to remember that you need friends everywhere. Teachers and students not directly associated with your program can benefit from it, too.

Avoid policies by fiat, and try hard not to make the program's success entirely dependent on one individual. Establish a target population, geographic range, and facilities after you have decided the size, direction, and needs of your program. Some of these decisions necessarily must be pragmatic and political.

In "Starting a Gifted Program" (Boston 1975, p. 26), several helpful lists are provided, one of which is cited in full here:

Principals, school board members, and superintendents will want to have the answers to the following questions:

1. Why do we need this program?
2. Why do we need this program now?
3. What does this program offer that is not being provided by existing programs?
4. How will this program fit into the rest of the instructional system?
5. How will present teaching schedules be affected?
6. Are we talking about a one shot deal, or something that has staying power?
7. How much will it cost?
8. Who is going to pay for it?
9. Who is behind it and what are their reasons for favoring it?
10. "But . . ." And here will follow a host of objections which will need to be anticipated as much as possible but which, if not, are at least answerable. . . .

Orenstein (1984) identified a set of strong organizational dimensions having a positive effect on gifted programs: Successful programs most often have formal structure with a built-in design for flexibility so that students and staff can take advantage of unique opportunities. Day-to-day activity is managed by a program coordinator or teacher who, being close to the students and instructional staff, has a good understanding of their needs. Budgetary control is often shared between this academic leader and a fiscal officer. Consultants frequently are used for assessing the needs of the program and for providing in-service assistance to the faculty, administration, and support staff. The staff is highly qualified and has had frequent and intensive in-service training. Operational details are important, and even the allocation of quality equipment and supplies contributes to the health of the program.

The creation of an oversight committee can provide different perspectives and contribute to the development and maintenance of a good program. Van Tassel-Baska (1981) recommends that such a committee meet frequently, have diverse membership, and not be dominated by program staff. The committee should include parents, leaders of local business and industry, educators from outside the program, and influential regional politicians as well as representatives from the program. Maintaining good rapport with all teachers helps to insure two-way cooperation and a better understanding of the special problems of the program and how they affect the rest of the school. The business representatives and politicians on the committee can play a special role in fiscal support.

Parents can be a powerful asset on such a committee, and it is helpful to have them organized as a separate group as well. Their input can be meaningful and valuable, and a climate that supports their interaction with the program staff and students should be cultivated. Specifically, and in their own best interest, parents can help to inform the community of the importance and quality of the program. A well-informed and supportive public is the best hedge against fiscal problems, and a basic message must constantly be reiterated: "Gifted students must have the same opportunity for academic and personal fulfillment as all other students. Anything less is a rebuff of a fundamental premise of public education." Some programs have also developed newsletters and resource directories to facilitate communication.

A task to be completed early is the recruitment and selection of a program leader, who should be appointed well before the program starts and who will take major responsibility for initiation activities. The lead time necessary depends on the size of the program. Even a small program will take several months; larger programs will need a year or more. The important thing to remember is that a committee cannot and should not undertake the operation of the program. Someone must be in charge. Authority can be delegated; responsibility cannot. A leader of the highest quality who can put his or her personal stamp on the program should be recruited actively. The rest of the staffing needs should be satisfied by matching the interests and abilities of persons with the tasks to be performed, not with roles. Try to identify the staff early and give them opportunities for input into all aspects of the program.

Counseling is an important and often overlooked aspect of operating a program for gifted students. They need academic and social assistance beyond that provided informally by classroom teachers. These students need help in making academic and career decisions, but they also need assistance related to their special ability—preparation, for example, to interact with a society that generally undervalues and sometimes even berates intelligence and academic interests. Van Tassel-Baska (1981) suggests also that all staff members and counselors meet frequently with small groups of students to discuss their problems.

Program policies should be formulated early and communicated to students and parents at their orientation to the program. In particular, the policy for removing students from the program, by staff or parents, should be made clear. Of course, provisions should be made to try diligently to avoid such a situation, but it is unreasonable to believe it will not happen. If parents and students have a reasonable expectation of the program, they will be less overwhelmed when they participate. It is also worth noting that program staff need similar information about expectations. They, too, can become overworked and overwhelmed, and special provisions should be planned to insure their satisfaction and comfort as well.

Design from the outset a component for evaluation, continual refinement, and modification of your program. Parents, teachers, and students, as well as those outside the program, should all be heard. New needs and ideas should be addressed for they can breathe vigor and vitality into everyone associated with the program. Furthermore, the documentation of the success of your program will tell the world what those intimately associated with the program already know. Everyone needs some good publicity. The staff and students in your program work hard and long, and they deserve public recognition of their achievements.

Assessments should allow for anecdotal reports. It is heartening for everyone to read, "We notice a difference in our son since he began the program . . . he is now excited, challenged, and again motivated to express himself and excel in areas where he has special talents and abilities" (Schimpfhauser 1986). Ongoing and well-designed evaluation is also necessary for grants and fiscal support and to answer critics.

A most important concern for any program for the gifted is funding. Without initial and continuing fiscal support *above the normal cost per student*, too little can be accomplished. Mitchell (1981) reported that almost all of the districts and states that she surveyed provided funding for education of the gifted student at higher-than-normal rates. Here are some samples of increments from Mitchell's findings: California $250, Connecticut $500 (by 1988), Idaho $657, Washington $255, and North Carolina $324. At the upper end, the 1983 budget for the North Carolina School of Science and Mathematics was $3.3 million for 400 students (Lyons 1983, p. 52).

Investigate the funding sources provided for your school and determine how you might tap them. This may be unfamiliar and unfriendly territory, but it needs to be studied carefully and with the utmost political scrutiny. In 1978 the United States Congress passed the "Gifted and Talented Children's Act," Public Law 95–561 Part A Title IX of the Elementary and Secondary Education Act (Alexander and Muia 1982). Unfortunately, this bill is no longer in effect, but new initiatives before Congress at the time of this writing may reinstate at least some of these funds; attention to this kind of activity is worth the energy of concerned educators.

Local funding sources such as businesses and industry, civic organizations, and political, philanthropic, and social organizations can provide not only program support but visibility and community awareness as well. A fact sheet from the U.S. Department of Education entitled "Finding Funds For Gifted Programs" (1978) suggests sources of information and strategies in your search for financial support.

One final suggestion comes again from Boston (1975, p. 27):

Power. Face it. Starting a gifted program requires changing the status quo. Concentrate energy, find a lever and a fulcrum, and lean. There are many sources of power—in persons, in information, in the ability to form and direct opinion and the energy of others, in the ability to deliver what someone else wants or needs, and in the ability to make advantageous trade offs, i.e., turning a "win-lose" situation into a "win-win" situation. The organizational component does not mean power grubbing, but it does mean taking the power issue seriously and unapologetically.

2
Guidelines

THE ISSUES discussed in chapter 1 make it clear that designing and operating programs for the gifted will present many exciting challenges and carry serious concomitant responsibilities. In this section we suggest guidelines to be considered by those associated with such programs.

Sixteen Essential Components of Programs for the Gifted

Any program for mathematically talented students should be expected to measure up on the following essential components:

Good Mathematics

The mathematical content of the program must be of a high quality, not just a collection of games, tricks, puzzles, or isolated topics. The curriculum should be qualitatively different from the regular program, not just an accelerated version of it. These differences should be reflected in the difficulty, cognitive level, breadth, and depth of the curriculum.

Sound Pedagogy

Teaching techniques must involve sound methodology and must be specifically appropriate for gifted and talented learners. Students should have specific instruction in the content they are learning. Teachers should encourage their gifted students to use their individual strengths to maximum advantage, and instruction should reflect the unique capabilities of the bright students. Instruction for the gifted should be delivered in a manner that would be inappropriate, even at a slower pace, for less able pupils. Teachers and learners should have significant interactions on a regular basis, and teachers should teach, motivate, and guide students, not just provide record-keeping and clerical services.

Teacher Competence

Teachers in programs for gifted mathematics students must be highly competent in mathematics, pedagogy, and educational psychology.

Higher-Order Thinking Skills

Programs for the mathematically gifted and talented must nurture high-level thinking processes. Students should work on open-ended investigations that stress higher-order thinking skills, and the entire content of the curriculum should be taught in a manner that demands application of such processes.

Applications and Problem Solving

Problem solving must be the major focus of instruction. Efforts must be made to include applications of mathematics to real-world situations as well as the examination of standard topics in greater depth.

Communication Skills

The ability to communicate is essential in learning mathematics. Students should be expected to read, write, listen, speak, and think mathematically. There also should be a heightened level of communication both between teacher and students and among the students themselves during the instructional process. The students should be expected to exhibit an appropriate degree of precision in their communication about mathematics.

Study Skills and Work Habits

Mathematics, with its unique content characteristics, provides an effective vehicle for developing study skills and work habits. These skills should include reading, note taking, studying for tests, problem solving, general work organization, and task commitment.

Individual Differences

The program should expect and accept a broad spectrum of differences in students even though they all have been identified as gifted and talented. This means not only variation in mathematical interests and abilities but also a wide range of activities outside the classroom. Gifted students typically have a broad spectrum of exceptional abilities, and their school programs should allow for and support their participation in extracurricular activities, sports, and clubs. They need opportunities to explore the many facets of their interests and personalities, friends their own age, and the appreciation that, although they are different, they are not "weird." Bright children are still interested in children's activities.

Encouragement of Creativity

The program must provide opportunities for students to explore mathematical ideas in a creative fashion. Students should be encouraged to experiment, explore, conjecture, and even guess.

Learning Resources

Gifted students need frequent and imaginative use of manipulative materials and other instructional aids. All programs need to employ a wide variety of learning resources including, but not limited to, texts, calculators and computers, television, other audiovisual materials, concrete manipulatives, and resource persons.

Integration of Content

Mathematics should be related to other content areas of the school program. This should happen in the mathematics class as well as in other subject areas.

Planning and Development

The total program must be well planned and coordinated. It should be developmental in nature and should foster the realization of undiscovered potential. Planning should include careful definition, sound identification criteria, and careful selection procedures, as well as the flexibility to modify the program as needs arise.

Evaluation

Ongoing evaluation both of student progress and of program effectiveness should lead to further improvements. Evaluation should be conducted regularly using a variety of approaches.

Student Concerns

The work of gifted students should be taken seriously. Teachers and program directors must listen to individual students' concerns and needs, and students in the special program should be allowed to interact with all dimensions of school activities. The program for the gifted must protect them from social isolation.

Mobility

Programs should be flexible enough to allow students to move without prejudice in or out of the program as their needs change. Decisions to add or discontinue a student from the program should, however, be made carefully and with input from school personnel, parents, and the pupils themselves.

Status

Status and prestige should be associated with the program. Both students and teachers should be given several and varied opportunities for recognition that extends beyond regular classroom activity.

4. Mark and Margie ran a 100-meter race, which Margie won by crossing the finish line while Mark was at the 95-meter point. In their next race, Margie gave Mark a 5-meter handicap by starting her race five meters behind the starting line. In the second race each ran at a constant speed exactly equal to the speed in the first race. Who won the second race?

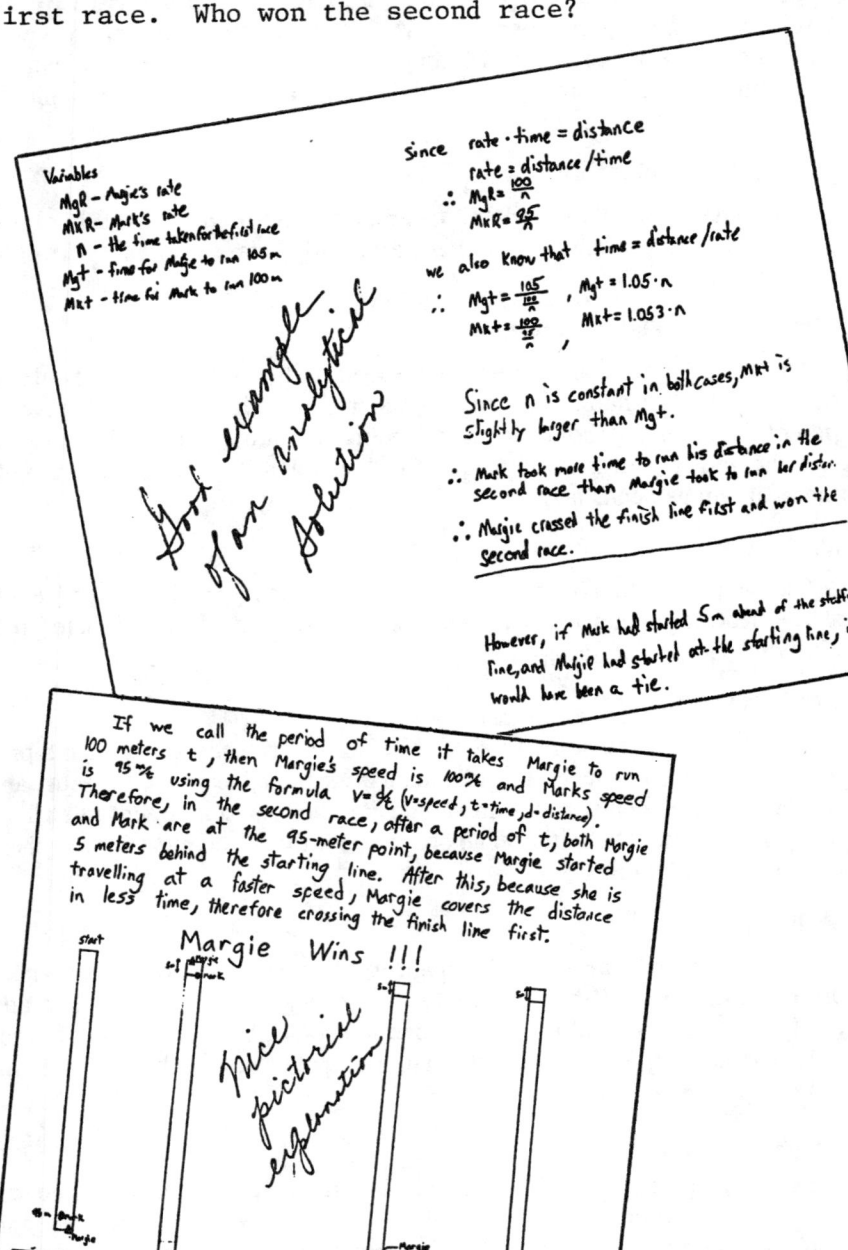

Elementary School Programs

Providing for gifted mathematics students at the elementary school level poses a special set of concerns and challenges. Major considerations are (1) How can we identify elementary school pupils who are gifted mathematically? (2) How can we modify the curriculum to provide for these children? (3) How can we provide support to elementary school teachers to enable them to provide appropriate opportunities for their gifted pupils?

Identifying Mathematically Gifted Children

Despite the recent attention to developing problem-solving skills in school mathematics programs, too often the dominant emphasis in both teaching and testing in the elementary school continues to be the development and practice of basic computational skills. The child who shows excellent facility with computational algorithms excels in school arithmetic and is the child most likely to be labeled "gifted" mathematically. Although children who are mathematically gifted usually do have good computational skills, such a method of identification is far too limiting and inconclusive. A child may have excellent skills in performing rote arithmetic computation but may be very limited in the ability to apply such skills in a problem-solving situation. Similarly, a child with average or less-than-average computational skills may have exceptional ability for finding strategies in game situations or for solving problems using geometric or spatial means or some nonstandard approach.

In order to attempt the identification of mathematically gifted elementary pupils, programs must include frequent opportunities to investigate varied mathematical topics and to use higher-level thinking skills in problem-solving situations so that gifted children can demonstrate their special abilities.

These special abilities can appear in many different forms. The young child who is gifted mathematically often demonstrates many of the following characteristics:

- Has a tendency to choose to do mathematics when presented with a choice of activities
- Masters typical content more quickly and at an earlier age than his or her classmates
- Often skips steps in problem solving and may solve problems in unexpected ways
- Is more willing and capable of doing problems abstractly; often prefers not to use concrete aids
- Enjoys and is successful at looking for patterns and relationships and attempts to explain them
- Concentrates for long periods of time on a problem that he or she finds interesting

- Has exceptional mathematical reasoning ability and memory
- Is more likely to see relationships between a new problem and problems previously solved; enjoys posing original problems
- Is capable of more independent, self-directed activities
- Enjoys the challenge of mathematical puzzles and games

Although a child may not demonstrate all the characteristics above, they are indicators of the traits usually observed in mathematically gifted children. Careful observation is the most promising means of identifying individuals who possess such characteristics; standardized tests offer minimal help because their focus is so limited.

Modifying the Curriculum

An elementary school teacher may have several pupils who fit the description of "high performers" in mathematics, but there is likely to be a great deal of variation even among these children, from the high achiever to the truly precocious child. How, then, can teachers organize their mathematics programs to provide for the high performers? Some typical practices include the following:

1. *Teaching the class as one group* but allowing such students to work on their own, either by progressing through the text at their own speed or by using other, more appropriate learning materials. Another variation is to teach a lesson to the whole class, structuring the instruction in such a way that both the in-class activities and the follow-up assignments are differentiated according to different levels of student ability. Such approaches may work, provided the students have opportunities to interact with the teacher and with each other. Without such interaction and regular feedback, however, the student is prone to racing through the book with minimal comprehension or to considering the activity merely busy work to be completed while the rest of the class is involved with the lesson.
2. *Grouping within the class* (as is usually done for reading instruction). This approach places a heavy burden on the teacher for planning different activities for each group every day. However, the approach provides all students with opportunities for regular interaction with the teacher and with peers.
3. *Grouping across grade levels*. In some schools high achievers in each class at a particular grade level are grouped together as a separate class for mathematics instruction. Although this makes it easier for the teacher to provide appropriate instruction, the approach poses many important questions: What criteria will be used to select the children? Will those criteria go beyond performance on standardized tests, yet be clear and defensible? Will all classes at a grade level have mathematics at the same time of day and for the same amount of time? What will the children be

missing when they are taken out of their classes? Is there a teacher available with sufficiently strong preparation in mathematics to teach the class effectively? Any grouping procedure also needs to be flexible enough to allow for changes based on ongoing information, and grouping does not relieve the teacher of the necessity to individualize instruction. Even a homogeneous group of high achievers in mathematics will include a wide variation in ability.

Regardless of whether a teacher is attempting to provide for a whole class or for a few children within a regular class, when selecting the mathematics content the teacher will be faced with the classic dilemma of acceleration versus enrichment. Should a bright child be allowed to go on to the mathematics text for the next grade? Or should the child be exposed to additional topics at the present grade level? Often the elementary school teacher faces additional pressure from parents who indicate that their child has already mastered the basic algorithms and should be working at a higher grade level.

Simply going on to the mathematics book for the next grade can be a great disservice to the child. It may appeal to parents who are eager to have the school recognize and provide for their child's talents, but it can be very shortsighted in the long run, especially if the gifted child is deprived of instruction that is specifically designed for his or her unique talents and needs.

Elementary school mathematics programs at every grade level contain more mathematics content than most teachers are able to cover in a school year. When the emphasis is to complete the book as quickly as possible, too often what is stressed is the development of skills with computational algorithms while less attention is paid to other areas of the curriculum such as geometry, measurement, probability, or statistics. Valuable opportunities to pursue these highly motivating and important areas and to develop essential estimation skills and broad problem solving abilities through activity-oriented investigations are lost. Since gifted children are able to deal more abstractly with mathematics, they often are deprived of the opportunity to experience regular use of concrete models to make connections between practical problems or applications and abstract mathematical ideas.

Several dangerous outcomes can result from too narrow coverage of topics in the elementary curriculum. The child may develop proficiency with arithmetic algorithms but only a shallow understanding of those procedures. Similarly, the child may perform well on standardized tests but at the same time possess an inadequate foundation of experiences with many areas of mathematics, which leads to serious difficulties in subsequent courses. Furthermore, because of the shallowness of their experiences in mathematics, many gifted students turn away from the subject in pursuit of more interesting fields.

A much more productive approach to providing for mathematically gifted elementary school children is to combine the best elements of both acceleration and enrichment. Although the children will not necessarily be using the text of the next grade, they will be learning concepts and skills that are included at higher grade levels. Such enrichment can take the form of investigating topics normally studied in the child's respective grade but in a deeper or more comprehensive way, studying topics that are not part of the normal school curriculum, or beginning to explore topics from higher grades.

Fortunately, an ever-increasing number of useful sources of enrichment materials is available to the teacher. Most recently published elementary textbook series include enrichment activities for more advanced students. These take the form of optional topics or extensions to the regular lessons. A few less widely used elementary mathematics programs intended for a general audience have special features that also make them valuable for gifted programs. An example is the *Comprehensive School Mathematics Program* (Heidema et al. 1978–86). Throughout that program there is a problem-solving focus with regular emphasis on developing higher-level thinking skills and proceeding from real situations to abstract mathematics. The program extends the usual curriculum to include such topics as probability, statistics, logic, combinatorics, affine and taxicab geometries, and network analysis. Special problems and differentiated activities provide additional challenge for gifted students.

Another less traditional program is *Developing Mathematical Processes* (1975). This curriculum uses activity-oriented measurement experiences as a vehicle for problem solving. The ninety units of instruction that compose the program are intended to allow for alternative organizational approaches and individualization. Unlike most elementary school programs, measurement and geometry play a dominant role, and emphasis throughout the program is placed on using higher-order thinking, on problem solving using concrete materials, and on the careful development of mathematical concepts and language.

Other commercially available enrichment materials include instructional guides, student activity books, task cards, computer software, and videotapes. Although these materials can be invaluable in providing enrichment activities for the gifted, it is imperative that they not become unrelated appendages to a basic program employed to keep gifted children busy. Enrichment materials must be selected carefully for their value as an integral part of a well-planned mathematics program. The focus of enrichment always should be to build a foundation for further study and to generate enthusiasm for, and understanding of, mathematics.

Renzulli (1977) offers an enrichment triad model that can be adapted to provide useful guidelines for the elementary school teacher in planning activities for mathematically gifted children. The three types of enrichment included in Renzulli's model are (1) general exploratory activities to stimulate interest, (2) group training activities, and (3) individual or small-group investigations of real problems. The following is an application of the model for elementary school mathematics:

Type I Enrichment: Exploratory Activities. This category of enrichment includes activities intended to expose children to new topics and experiences. These can take place in an interest center in the classroom where the children are free to choose among activities and to explore on their own without being required to prepare a formal summary or report. Many materials and activities are especially useful for Type I enrichment. Some examples include the following:

- Commercially available manipulative materials such as pattern blocks, Cuisenaire rods, and geoboards. These materials are designed to lead to the discovery of many fundamental patterns and relationships. At this stage the children need to use them freely; task cards and activity books should be available to suggest possible activities.
- Number puzzles and games, which can be highly motivating. Classic examples include Nim, the Tower of Hanoi, and magic squares. Hundreds of such games and puzzles are available.
- Geometric and spatial activities, which are extremely important but which usually are not emphasized in the typical program. Activities such as solving tangram and pentomino puzzles, making three-dimensional geometric models, curve stitching and creating line designs, and making scale models provide these important experiences.
- Calculators. Experimenting with hand calculators can lead to fascinating Type I experiences when, on their own, children explore the basic operations and their relationships to one another, number patterns, computational shortcuts, and more.
- Computers. An ever-increasing and improving selection of software is especially appropriate for introducing children to computers. Computer games and simulations offer an informal, motivating introduction to the possibilities of using technology.
- General problem-solving activities. Many kinds of task cards and activity books offer hundreds of interesting and varied problems from which children can choose.

A selection of reading books on mathematical topics can provide still more Type I enrichment experiences. Many books are available dealing with such topics as symmetry, area, volume, polyhedra, probability, graphs, topology, and so on. They can open up new horizons and suggest further investigations that may interest the child.

Type II Enrichment: Group Training Activities. These activities are extensions of the Type I explorations and are intended to focus on the development of important thinking processes and key mathematical concepts.

- Using manipulative materials in a more structured, teacher-directed way in order to focus on the discovery of key concepts. For example, a series of formal activities with pattern blocks to examine how basic geometric

shapes "grow" can lead to key ideas about similarity, ratio, and proportion. Sequenced activities with geoboards can result in careful development of the distinction between area and perimeter and of how each changes as the figure is transformed.
- Looking at winning strategies for games and alternative solutions to puzzles, or analyzing what makes a number trick work.
- Organizing geometric and spatial activities so that they lead to careful attention to such important ideas as the different kinds of symmetry and ways of identifying and creating symmetric shapes; the effects of different kinds of transformations on geometric figures; ways of representing three-dimensional figures in two-dimensional perspective; and basic applications of linear, area, and volume relationships using regular and irregular shapes.
- Instruction in the effective use of calculators, such as specific techniques for simplifying a series of operations, using the memory capability, and representing fractions as decimals.
- Instruction in the effective use of computers. Students can examine successful strategies for computer simulations or games. Learning a computer language such as Logo presents especially good opportunities to engage in meaningful problem solving at a very young age. Using simple graphics, children learn to sequence instruction to produce a definite result, break a problem into smaller parts, use subprograms to simplify a project, look for alternative solution approaches, and more.
- Teaching specific problem-solving skills. This includes looking at the processes used in effective problem solving and developing a deliberate focus on such heuristics as making diagrams, working backwards, estimating, using trial and error, organizing data in tables or graphs, and looking for patterns and relationships.
- Organizing experiments that lead to the discovery of basic probability concepts. This can include performing experiments with concrete materials or using computers to simulate such situations.
- Planning experiences in collecting different kinds of data and organizing those data into convenient formats including stem and leaf plots, histograms, pictographs, and bar, line, and circle graphs. Such activities would also include examining the similarities, differences, and relative usefulness of these graphs in various situations.

These are only examples of the many kinds of Type II enrichment experiences of value for elementary school programs. They can be thought of as extensions of Type I activities, but these are carefully structured to lead to the development of important thinking skills and key mathematical ideas. They provide a foundation for Type III experiences.

Type III Enrichment: Investigating Real Problems. This type of activity helps to identify and challenge the most gifted students. The essence of Type III enrichment is having children investigate complex problems that are interesting to them. They become problem finders and problem posers in these self-initiated investigations, and they have an opportunity to present their findings to a real audience.

Activities of this type are less teacher directed, but the teacher must provide opportunities for them to occur. For example, when investigating a problem like the classic pentomino problem (In how many different ways can you arrange five congruent squares so that they are joined along whole sides?), some children will naturally experiment with variations of the task. The teacher can encourage such experimentation with "What if?" questions that stimulate problem creation: What if there were only four squares? Six squares? What if instead of squares we used equilateral triangles or rectangles or regular hexagons? What if we use five cubes and think of the problem in three dimensions?

The teacher must be willing to ask and encourage open-ended questions such as these and to provide time for their investigation. A good problem is rarely answered in a few minutes; it usually requires a much more extended time period. The student should expect to return to the problem many times and to discuss strategies and approaches, alternative solutions, variations of the original problem, and tentative outcomes.

Opportunities for Type III enrichment can come from the other types discussed earlier. Gifted students enjoy opportunities to undertake such activities as these:

- Inventing their own problems, possibly as an extension of activities used to develop problem solving skills
- Modifying a game or puzzle by changing the rules or the way the game is organized
- Inventing new algorithms or procedures for a particular computation or problem and proving that their methods work
- Writing their own computer programs for a purpose that interests them

The self-initiation of an investigation, seeing the project through to conclusion, and sharing the results with others are the major components of Type III enrichment. The teacher plays a critical role in each of these components by—

- stimulating and encouraging questions that will lead to independent investigation;
- providing time, resources, encouragement, and guidance when needed;
- making it possible for the student to share the results of the investigation with other students, teachers, or interested persons outside the class.

Clearly the three types of enrichment described above are not intended solely for gifted students. The first two types provide valuable extensions to the curriculum that are important for all students. The third type is most appropriate for the gifted, however, since these activities require the higher-order thinking processes and capability for independent work that characterize the talented.

Careful, deliberate planning by the teacher is important in any enrichment program. But two cautions are in order when anticipating such activities: First, gifted children are sometimes reluctant to use manipulative materials and prefer dealing with mathematics abstractly. Although they should have opportunities for abstract thinking, they also need to discover mathematical relationships and structures from work with concrete materials and practical situations. Such experiences help them to appreciate the power of particular mathematical abstractions inherent in different applications and to begin to realize the power of mathematical modeling and simulation—two crucial contemporary uses of mathematics.

The second concern is that enrichment experiences must not replace more standard content to the point where they become the entire curriculum for the gifted. Children still must learn important concepts and must develop, practice, and maintain appropriate computational skills. Instruction in those topics and skills can proceed at a faster pace with the talented, and many creative approaches can be used to make drill more enjoyable or to embed practice within enrichment activities, but during exposure to enrichment experiences gifted children still must be responsible for mastering essential mathematical content.

Supporting Elementary School Teachers

Programs for gifted children based on activities like those described above are highly dependent on the teacher. In elementary schools this poses a significant problem, since elementary teachers are responsible for many subjects. Often they have minimal background in mathematics and may even dislike and fear the subject. Such teachers are only comfortable teaching mathematics in a very prescribed, algorithmic fashion, and they are unlikely to welcome the types of activities suggested earlier, which require background in mathematics, a spirit of inquiry, enthusiasm for the subject, and problem-solving skills. They may prefer the safer route to providing for the gifted by using a higher grade textbook to teach more advanced computation.

There is a very real need to identify, as instructors for the gifted, teachers in the elementary school who have strong backgrounds in mathematics and who enjoy the subject. They must then be supported with curriculum resources, opportunities to extend their knowledge and skills, and planning time to organize well-conceived programs. Eventually they can provide leadership for the entire school in providing opportunities for the gifted.

But since the majority of mathematically gifted children still remain in regular classes, it is equally important for schools to provide more opportunities for all elementary teachers to develop their backgrounds in mathematics and to experience learning mathematics using the same activity-based problem-solving approach advocated for children. Brief workshops or in-service presentations will not suffice to achieve this goal; extended, ongoing opportunities will be necessary within a nonthreatening, supportive environment such as a teacher center.

Elementary school teachers also must become aware of the many, varied curriculum resources available for enriching mathematics, and they must have opportunities to use such resources and to have a voice in the selection and evaluation of enrichment materials for their classrooms.

Making these opportunities possible is probably easiest in a system that has a coordinator of elementary school mathematics and an ongoing program of staff development. Without a commitment to providing adequate support for teachers, it will be impossible to implement on a broad scale the enrichment programs that mathematically gifted elementary school children need and deserve.

Middle, Junior, and Senior High School Programs

There are special concerns at each of the levels from middle school through senior high school, but there are also many common features that mathematics programs should have if they are to prepare gifted students for future success. The remarks that follow are intended to be neither prescriptive nor a design for any particular program; rather they represent the thrust and character of the ideas and attitudes that should underlie all programs.

When students leave elementary school they are at a time in their lives when the world is opening for them, and they begin to see the magnitude of the choices they can make. These are years when many students become disillusioned with school. They can be easily distracted, even disruptive, or they may just drop out of mathematics programs altogether. They feel the pressure of their peers to conform, a pressure that often implies holding academic standards in disregard. Young women and minorities, especially, need substantial positive reinforcement, concern, and support from respected adults.

Bright students may, as they move to a larger school, encounter for the first time students whose abilities equal or exceed their own. They will need help both in adjusting to this new environment and in meeting teacher expectations that are different from those of their elementary school teachers. They may have inadvertently developed poor work habits as a result of instruction aimed below their abilities, and consequently they may come to expect that they do not need to pay attention, do homework, or study for tests. They are usually unaccustomed to the postponed gratification associated with the continuing, persistent activity necessary for the serious study of any subject. Without the watchful eye of a nurturing teacher who helps them to organize their time, ideas, and even materials, they may be cast adrift.

Parents of these students very likely have come to expect their children to complete their schoolwork with excellence, and they may not be prepared to consider the new demands on their children's resources. Time must be taken, with students and parents, to address work habits and study skills. Note taking, homework, and studying are acquired skills, as is the ability to organize time and energy. If we want students to participate with pleasure in the range of activities they may choose, we must address directly the acquisition of these now necessary skills.

We must help them to see mathematics as a reasonable, interesting, and enjoyable conduit for their time, energy, and abilities. As Krutetskii (1969) pointed out, substantial mathematical ability is neither necessary nor sufficient for keen interest in the subject. Students at this age are developing their interests, and although we want them eventually to be independent of teachers and school and to have autonomy in choosing the topics they wish to study, at this formative stage we owe them guidance toward a wider perspective. Children must be nudged to go beyond previously developed interests and not to remain strictly within their current styles of mathematical operation. They need to develop forms of mathematical thought not previously in their repertoire (Ridge and Renzulli 1981). To that end we should be cautious of what Renzulli describes as the "revolving door" model (Renzulli, Reis, and Smith 1981) that encourages students to participate only when they have the inclination. Above all they need—

- something interesting and worthwhile to think about;
- something interesting and worthwhile to do;
- the nurturance of a creative, knowledgeable, and sensitive teacher.

Some gifted students are markedly creative and have enormous potential, but by this time in their academic careers they may be deficient in conventional skills such as penmanship, grammar, and basic computation. Providing sufficient drill and remediation may be necessary to lay a foundation of skills on which they may build. Enrichment and acceleration must be carefully integrated into gifted programs. So, too, must drill, practice, and review.

Students can participate effectively in investigations, and they should have the opportunity to study in a discovery format that has a quality of open-endedness. Maker (1982) insists that students be given freedom of choice for investigations and group activities, and she views such choices as critical to the students' learning of social and leadership skills. The formal and informal communication necessary in group work can also help students to recognize the value of good English and good notation. But teachers must beware. Students also need guidance and reasonable access to appropriate materials, and teachers must retain an unobtrusive but diligent managerial role. True discovery takes time and involves risks. Left totally on their own, students may "discover" the wrong thing or make no discovery at all.

Aiken (1973) identified two basic mathematical types of mind: the slower, logical, formal type typified by analysts and the quicker, intuitive type associated with geometers. Krutetskii (1976) observed, in addition to those two, a third type that harmonizes both analytic and geometric approaches. Group work should provide an opportunity for students to recognize such differences in styles of thinking and help them to appreciate their own thinking and the value of alternative styles. Students should be encouraged to become familiar and comfortable with different styles, even ones that differ from their natural preferences. Clear thinking and effective communication should be expected of everyone. Projects and reports should be developed and, when necessary, reorganized or rewritten to a polished completion.

Contests, problems, and what has come to be known as recreational mathematics can supply a rich source of good ideas for study. The books of Martin Gardner, Sam Loyd, and especially the *Elements of Mathematics (EM) Book B Problem Book* (Engel et al. 1975) all provide opportunities for serious problem solving and creativity by allowing students to answer nontrivial and nonroutine problems requiring no special prerequisites beyond basic mathematical skills. Problems in the EM book have an indicated level of difficulty from 1 (lowest) to 5 (highest). Figure 2 shows a sample of a level 2 problem and an eighth-grade student's solution.

Success at contests requires not only a good deal of special preparation but also accuracy and speed under pressure. Furthermore, many national and international contests tend to include problems requiring a knowledge of esoteric or specialized theorems, especially ones from geometry and number theory. But competitions do appeal to many young people and can provide a vehicle for recognition as well as a source of some interesting problems to solve. Contest questions can also be used as generators for creative problem posing, for developing skill in generalization, or for problems that require essay type responses.

Ridge and Renzulli (1981, p. 229) noted that "gifted students have a penchant for abstract thought, often to the virtual exclusion of the practical," and they argued instead for considerable exposure to practical considerations leading to mathematical modeling and simulation. But there is a fallacy in their admonition (1981, p. 230) that mathematically gifted students who are encouraged to deepen a consuming interest in pure mathematics to the utter exclusion of any knowledge of applied mathematics can cause them to run the distinct risk of joining the glut of Ph.D.'s with nowhere to go but behind the wheel of a taxicab. First, recent evidence (David 1985) suggests that there is and will continue to be a shortage of Ph.D.'s in mathematics and the natural sciences. But a more important refutation of Ridge and Renzulli is that much of what is proposed to be "real-world problem solving" is very often contrived and, quite frankly, boring. Promoting a problem as "real world" is insufficient motivation for many students. What they want are problems that are interesting. Additionally, we should capitalize on the creativity and imagination of these children before they have had fantasy and playfulness schooled from them.

In *The Art of Problem Posing*, Brown and Walter (1983) describe a classroom style of investigation and communication that can be of the highest order. They effectively support their claim that no amount of "merely understanding" can take place without problem generation. Also outlined is a classroom strategy in which students act as authors and editorial board members in the production of a class journal that is developed over the course of a semester or year. Students write their work for public scrutiny and evaluate the work of others in a manner that is normally reserved for professionals.

Students should be encouraged to see mathematics in settings outside of their classrooms and to begin to acquire the habits of a professional. They can read articles in professional journals such as the *Student Math Notes, Arithmetic Teacher, Mathematics Teacher, Mathematics Magazine*, the *American Mathematical Monthly*, and *School Science and Mathematics*, as well as in popular publications like *Discover, Scientific American, Omni*, and *Games*, and they can take an active role by submitting solutions to published problems or writing letters to the editor. They should have access to books other than textbooks, such as, for example, those of the Mathematical Association of America's *New Mathematical Library* series.

It is important to remember that these gifted students are children and have children's interests in games, fairs, and contests. Similarly, while they have the ability to abstract and formalize, they also need opportunities to use manipulatives and to play. Even at a Piagetian formal operational level, concrete objects and drawings are not only useful, but necessary. Strategy games can also help students develop concepts of proof and an appreciation of formal structure. Here the rules of the game are the axioms, the winning strategies the theorems.

Gifted students need access to calculators and computers and sufficient instruction to be able to use them effectively. It is, however, inappropriate to substitute courses in computer science or programming for mathematics, even though such courses may contain ideas of particular appeal. Student work with calculators and computers warrants time in the mathematics classroom only when it contributes directly to the mathematics curriculum. There is, for example, strong mathematics and problem solving content in such aspects of computer science as discrete structures, artificial intelligence, algorithm design and information representation, and students should be made to see the role these play in contemporary mathematics. Computers can also provide unique opportunities through such specialized software as the *Geometric Supposer* (Schwartz and Yerushalmy 1985) or *Green Globs* (Dugdale and Kibbey 1986) or books like *Turtle Geometry* (Abelson and diSessa 1980).

One of the most important qualities of secondary school programs for gifted students is the development of proof. In particular, proof should be considered in the context of a working mathematician: an argument that convinces. Figure 3 gives an example of such an argument from a middle school student (Rising and Harkin 1978).

As students progress through school they should develop greater logic and formalism, which will lead to the style of proof exhibited by the ninth-grade student whose work is shown in figure 4 (Kaufman, Fitzgerald, and Harpel 1981).

The enrichment activities associated with commercial textbooks are insufficient for gifted students, as is mere acceleration. Two major curriculum projects, the Secondary School Mathematics Curriculum Improvement Study's (SSMCIS) *Unified Modern Mathematics* (Fehr et al. 1968–1972) and the Comprehensive School Mathematics Project's *Elements of Mathematics* (EM) series (Martin 1970–1983), provide complete programs for secondary students. Both these curricula are integrated approaches that represent a significant departure from standard programs and may require considerable organizational changes for implementation in most schools, but educators should seriously consider all or parts of them.

When students achieve in mathematics above their standard grade levels, they should get appropriate credit, and their official transcripts should reflect the level of their achievement. If high school graduation requirements are based on Carnegie units or some other system of credit accrual, then middle or junior high school work may warrant senior high school credit; if senior high school work merits college credit, then it should be granted. This will require understandings and agreements among all levels of the educational continuum. Making those arrangements may be difficult and time-consuming, but students should not be penalized for their work by having to repeat courses or delay advancement. Similarly, students should not be penalized by a grading system that measures them by one standard and then compares their grades with those of students who have been measured by another. Average work in a gifted program does not warrant a C on a student transcript.

Many items of mathematical interest can be used over a wide range of grade levels, and a spiral curriculum in which students return to a topic previously studied most often leads not only to an increased depth of already studied ideas and an integration of various aspects of mathematics, but also to the kind of growth in sophistication that mathematical maturity warrants. Middle and junior high school topics include the following:

- Number theory
- Elementary, intermediate, and abstract algebra
- Probability
- Statistics
- Combinatorics
- Logic
- Synthetic, transformational, and analytic geometry

- Topology
- Topics from discrete mathematics such as elementary graph theory

Senior high school topics include the following:

- Combinatorics
- Non-Euclidean geometry
- Classical analysis
- Advanced, abstract, and linear algebra
- Topology and advanced topics in geometry
- Topics from discrete mathematics, such as algorithm analysis, graphs and trees, difference equations, linear programming, recursion, and matching
- topics from the theory of computation such as Turing machines

Finally, in planning for gifted students in these grades, the following considerations are important. For middle school and junior high school:

- Don't stress formalism too early. Build bridges from the concrete to the abstract. Games play a useful role here.
- Build positive attitudes toward mathematics and toward learning in general.
- Emphasize the long term implications of choices made at this time.
- Help students see real people using mathematics.
- Lay the groundwork for more formal work in later grades by dealing seriously with logic and proof.

For senior high school:

- Don't let student acceleration lead to years with no formal mathematics instruction at all.
- College textbooks and a lecture recitation format can be deadly. Whenever possible work with students as coinvestigators.
- Provide career guidance.
- Help students see the debate and inquiry typical of higher mathematics but usually never exhibited until the upper division undergraduate or graduate school level.
- Expect students to read and write mathematics as contemporary mathematicians do.
- Help students work on projects, extended topics, or individual investigations that require long-term commitment.
- Emphasize proof, structure, and understanding.
- Prepare students not only for calculus but for other branches of mathematics as well.

At all times, educators must remember that they are asking a great deal of young gifted students, and the students in turn should rightfully expect and receive a great deal in return.

Program Organization

The previous sections discussed many of the broad concerns and issues to be considered in developing programs for the gifted. We turn now to some specific considerations to be faced at different levels of organization within the K–12 educational system.

Providing for the Gifted in a Regular Class

Most teachers attempt to provide enrichment activities and extensions to the regular mathematics program in an effort to meet the needs of their most able students. Dealing with the exceptionally talented mathematics student within a regular class setting poses an even greater challenge and requires additional efforts. The following approaches suggest strategies the classroom teacher may employ in meeting that challenge.

- Be flexible about assignments. If the student can do the assigned work at a faster pace, or already knows the material being covered, it is unreasonable to require him or her to move at the same pace as the rest of the group. Ask yourself whether all the reinforcement exercises are really necessary for the student who has already demonstrated mastery of the given concept or skill. Discuss possible alternative assignments that are more motivating or productive for the student. This includes being willing to have the student work on problems that are different from those you might suggest.
- Seek problems and activities that admit of varied approaches and many levels of solution so that students can participate in a common task while experiencing success in individual ways.
- Obtain additional resources to supplement the basic program. Such resources can take many forms, including printed materials, access to computer facilities and challenging software, opportunities to interact with other adults who can serve as mentors and role models, and participation in mathematics contests and fairs.
- Allow gifted students to exercise appropriate leadership in pursuing special projects and investigations. While maintaining high expectations and accountability, give them opportunities to plan their own investigations and to decide how to allocate their time.
- Spend time with them in extra dialogue. Although gifted students generally have the ability to work independently, they need periodic guidance. Through discussions you can make it possible for students to describe their projects, and your questions can stimulate them to consider alternatives or extend their thinking.

- Provide opportunities for the students to communicate with others about their work. This can be done in many ways, such as informal sharing sessions, formal presentations to peers or other interested parties, written journals, and articles submitted for publication.
- Make a special effort to communicate with parents. This helps bring recognition for the student's work while affording parents the opportunity to share in the planning and to help relate in-school and out-of-school activities.

Teaching a Class of Gifted Students

Teaching an entire class of gifted students makes it possible for the teacher to modify the curriculum to meet the needs of the pupils. Some ways a teacher can do this are as follows:

- Recognize differences among students in the class. Even though all have been selected for their superior abilities, they are likely to be more different from each other than they are collectively from the "average" students. Very likely some will have high general intelligence and will have been exceptional achievers in school, but they may lack some of the natural insights and abilities characteristic of the very gifted in mathematics. Such students, who may have been stars in their regular classes, can be intimidated in the special class. Your sensitive intervention will be very important to their success in the new environment.
- Modify the mathematics content. Don't expect to use the same textbook or learning materials as with an average class, not even if you proceed at a faster pace. Mathematically gifted students should study mathematics in greater depth and at a higher level of abstraction. Locate and use materials that respect and provide for their abilities.
- Encourage active participation. Here again the teacher plays a critical role in establishing a classroom environment that encourages each student to contribute his or her ideas, facilitates student-to-student dialogue, and rewards clear verbalization of the processes used to solve problems rather than a mere statement of answers.
- Prepare for the unexpected. Gifted students often respond in highly original and unexpected ways and approach problems in a manner completely unanticipated by you. You must be willing to listen to such responses and, at times, to modify the lesson plan to take account of a student's suggestion.
- Pay special attention to how you pose questions. Questions should help students consider alternatives and extensions to the problem at hand. Especially important are open-ended questions like "How many different kinds of _____ can you find with the property that _____ ?" "How many different ways are there to _____ ?" "What if we change the

given conditions and assume instead that _____ ?" "When will it not be true that _____ ?"

Programming on a Schoolwide Basis

Despite the exemplary efforts of individual teachers, programming for the gifted requires schoolwide cooperation to establish an effective, coordinated program. In such efforts, the principal plays a leading role by providing the essential encouragement and support that teachers need in order to plan and implement programs within their classrooms, and by making possible the development of alternative organizational structures such as learning centers or a school within the school. Following are some of the ways in which the principal can exercise this leadership:

- Openly demonstrate genuine commitment to establishing and maintaining the program. Many of the problems associated with selecting students, involving teachers and parents in decision making, securing adequate funding, and evaluating outcomes are time-consuming and difficult to resolve. Administrative convenience must not be allowed to dictate policy at the expense of the needs of children.
- Recognize that honors classes or their equivalent constitute separate preparations for a teacher even though they may bear the same title as a regular course.
- Be certain to involve teachers and parent representatives in planning and decision making. Make them aware of the important issues involved, such as methods of identification and curriculum considerations; opportunities that have been developed elsewhere in and outside of the system; and recommendations of experts with experience in gifted education.
- Establish regular opportunities for communication. Teachers of classes for the gifted need to interact regularly to discuss their efforts, share ideas, compare activities and outcomes at various grade levels, and coordinate activities across classes. Interaction between teachers of gifted classes and those not teaching such classes also is important. Developing and maintaining the program must be a schoolwide commitment, not the sole responsibility of a small group or a single individual.
- Communicate beyond the staff. Keep parents informed about the program, clarifying how the program is organized and giving them opportunities for contributing to its development. Provide a communication link with the rest of the school district, making others aware of what the school is doing and bringing back to the school information about significant activities and opportunities elsewhere.

Coordinating Districtwide Efforts

Support for gifted programs can come from districtwide efforts in a number of ways. Examples of such involvement include the following:

- Coordinate the efforts of several schools: As schools share information about their programs, efforts should be made to assess what is done at different levels and to work cooperatively to coordinate developments at the elementary, middle, and secondary levels. District personnel also should monitor disparities of opportunities for gifted students in different schools.
- Provide support services to individual schools: It may be possible for the district to provide resources not available in individual schools, such as resource persons, curriculum materials, clerical assistance, and special equipment. The district also can facilitate communication among schools about local efforts for gifted students and can make schools aware of outside resources that may be available to schools, teachers, or students.
- Organize districtwide initiatives: By pooling the resources of the district, it may be possible to offer opportunities not available in individual schools. Examples are districtwide mathematics fairs and contests, special curriculum materials, and supplementary enrichment programs administered by the district. Where schools are too small to support a school within a school, the district might establish a magnet school to serve the gifted.

Making Multidistrict Provisions

Small school districts that are unable to support large-scale gifted programs can work instead with neighboring systems in a joint effort to develop resources for the gifted. Examples of cooperative initiatives are regional resource or learning centers and specialized schools that draw gifted students from all cooperating districts. Organizing such joint efforts will entail some important considerations that educators must be prepared to face as they plan cooperative programs:

- How will adequate communication be established? Joint efforts must be carefully planned so that each participating district has representation and ongoing involvement in decision making.
- If a special school is established, what will be the selection procedures? Will such procedures be used consistently? Will a quota system be used? Quota systems can pose significant problems, not the least of which is selection procedures that are modified to meet quotas.
- Where will the special school or resource center be located? Will it be convenient for all districts involved? What transportation arrangements will be needed?
- How will the facility or program be funded? Will it compete with existing facilities and programs? What new sources of funding are available?
- What problems can be anticipated as a result of removing students from their home schools? Will schools cooperate with the plan?

- How will the faculty be selected? Will a faculty quota system be used to assure representation from all participating districts?

These are realistic questions that districts must face in any attempt to coordinate efforts. Joint ventures can be successful in making available for gifted students opportunities that would not be possible within the individual school district.

Initiating and Managing the Program

Careful, deliberate planning is the key to success for the gifted education program both before and during its implementation. Program planners and managers should develop a checklist of tasks along with dates by which they are to be accomplished and identification of the persons to be responsible for each. We close this section with an example of such a checklist drawn from suggestions of the Minnesota Department of Education (1986).

A. Develop the program plan.
   1. Recognize the need for special programs.
   2. Establish a task force to be responsible for the planning activities listed below. Include administrators, teachers, parents, counselors, appropriate specialists, students, and alumni.
   3. Develop the program's philosophy, goals, and objectives.
   4. Develop a working definition of "gifted and talented."
   5. Assess currently available provisions for the gifted.
   6. Study alternate types of programs.
   7. Survey current literature on gifted education.
   8. Visit existing well-established programs.
   9. Relate proposed and existing program options to goals.
   10. Write the gifted education program plan to include the following components:
       a. Philosophy of gifted education
       b. Goals for the program and objectives for accomplishing those goals at each educational level
       c. Identification of the program coordinator
       d. Methods of identification based on the operational definition of giftedness
       e. A plan for involving parents, community members, teachers, and administrators
       f. Description of options to be offered at each grade level
       g. A staff development plan
       h. Program evaluation design
       i. Budget

      j. Dates for local approval and implementation
- B. Implement the program plan.
    1. Establish an advisory committee for gifted education with representation similar to that of the planning task force.
    2. Provide appropriate information to staff and community.
    3. Conduct staff development through workshops, program visits, conferences, and so on.
    4. Provide curriculum writing time.
    5. Design an evaluation plan and identify instruments to be used.
    6. Identify students for the program.
    7. Provide information to the parents of selected students.
    8. Implement the program as planned.
    9. Collect baseline data to use in evaluating the program.
- C. Evaluate the program according to the plan.
- D. Maintain and refine the program.
    1. Analyze results of the evaluation.
    2. Determine areas of need for refining the program.
    3. Implement modifications; make additions or deletions.
    4. Continue staff development activities.
    5. Assure continued maintenance and refinement.
    6. Assure ongoing evaluation.
- E. Expand the program.
    1. Analyze the results of the continuing evaluation.
    2. If warranted, follow previously developed planning procedures to design and implement the proposed expansion.
    3. Continue to evaluate, maintain, refine, and expand as needed.

3
A Closer Look

THROUGHOUT the country there are in operation numerous exemplary programs for gifted mathematics students. In this section we present a closer look at several examples that represent alternative approaches. These are intended not as models to be cloned but as illustrations of a range of successful strategies for expanding opportunities for the mathematically talented. Other examples could have been chosen, and readers are encouraged to seek out and visit successful programs in their local areas. Names and addresses of contact persons for the programs described below are included in the Appendix.

Special Schools

"Special High Schools: A Renaissance," a lengthy *Los Angeles Times* (1983) feature article, reported that special schools have been undergoing a rebirth, although the notion of special schools for especially talented students has been around for a long time. These schools typically have rigorous selection processes and draw from population centers large enough to make their programs viable. Some with small geographic boundaries are located in urban areas of dense population where students commute. Others encompass an entire state and provide complete residential programs. Their curricula have great diversity. A few have long-standing and impressive reputations.

Secondary Schools

The Boston Latin School, for example, was founded in 1635, and although it does not call itself a school for the gifted, its competitive selection process, rigorous standards, and more than 25 percent attrition make it clearly not a school for average students. Seventh through twelfth graders participate in a program having a unique pedagogical approach, especially by modern educational standards. Rote memorization, orderly progression of content, and repeated drills are standard classroom strategies, and "rhetoric and the classics foster the very skills in logic, analysis, and persuasion necessary to top flight lawyers, physicians and scientists" (Wernick 1985, p. 135). Students have many requirements and few electives.

The Bronx High School of Science offers an intensive academic program to its 3200 students, drawn from New York City. The primary teaching method is Socratic, in which questions are raised and challenges offered to students who must then obtain their own conclusions. Students are required to take four years of English, social studies, and science; three years of mathematics and a foreign language. Many unusual electives are offered, including astrophysics and laser optics, and students have opportunities to participate in full-scale experimental research (Wolkomir 1985). Students from this school consistently are National Merit Scholarship finalists and winners of national contests like the Westinghouse Talent Search and the American High School Mathematics Examination.

Recently, several states have undertaken large-scale projects in response to the need for special programs for talented students. The pioneer was North Carolina, whose governor and legislature in 1978 established the North Carolina School of Science and Mathematics (NCSSM), which has served as a model for the entire country. In 1984 NCSSM was brought under the administration of the University of North Carolina so that more opportunities for research and increased educational resources could be made available. NCSSM is a public residential high school for eleventh and twelfth graders. Amply funded by federal, state, and private sources, it has four hundred students and its own board of directors. Students from across the state are nominated for admission during their sophomore year. To qualify, students must meet a required standard on the Scholastic Aptitude Test and on a special local test designed to measure such qualities as abstract reasoning and creativity. Student credentials are then reviewed by a selection committee that accepts young people whom they judge to have high potential for academic success, interest in intellectual concerns, willingness to learn, inner discipline, capacity for independent thinking, ability to work well with others, and desire to serve the community.

The teaching and learning process includes independent study, tutorials, projects, and professional research internships. The curriculum emphasizes mathematics and science while not neglecting the humanities and physical education. Since the background of the students is diverse, mathematics courses range from elementary algebra to courses traditionally taken by college undergraduates; independent study arrangements also are available. Mathematics and computer science form a combined department that emphasizes problem solving and breadth of mathematical experience. The traditional sequence of analysis, multivariate and vector calculus, and differential equations is available, as are finite mathematics and a variety of computer science courses. A complete listing of mathematics courses is described by Davis and Frothingham (1985), and an information brochure is available on request from the school.

During their two years of residence, the students must complete twenty-three to twenty-four credits: three in science, two in mathematics, two in English, one in social science, one or two in a foreign language, one-half in physical education, and one and one-half in electives. Each student must also demonstrate computer literacy and fulfill two years of work service and one year of community service. The core curriculum is supplemented each year by special seminars and symposia. Students are encouraged to participate in problem-solving sessions and mathematics contests and to attend an annual lecture series presented on campus by mathematicians from universities and industries.

The faculty and staff of NCSSM are well qualified; each holds a master's degree, and about half hold a doctorate as well. Additionally, they are selected for their integrity, compassion, and dedication, which are necessary to make a residential high school program work. To supplement the thirty-eight regular faculty members, counselors, resident advisors, and administrators, the school employs invited interns who are temporarily reassigned from other schools in North Carolina to teach and develop new materials.

The school also has a broader challenge to serve the entire state through outreach programs. Faculty members regularly speak at statewide meetings, and the school conducts summer workshops for teachers both on its campus and at sites across the state. The Summer Ventures program allows six hundred high school juniors and seniors to participate in a five-week program of activities at sites across the state, thus providing a "less traditional, interdisciplinary look at scientific and mathematical processes" (Matros 1985). Students in the Ventures program divide their time about equally between mathematics and science activities. As with NCSSM, no tuition is charged, and room and board are provided without fee.

The success of NCSSM is evident in the number of national awards and scholarships won each year by its students, a success that is due to a unique combination of factors: a singular campus; dedicated faculty and staff; outstanding students; and academic, cocurricular, and residential programs tailored to their needs. Unique also is the close relationship NCSSM has established with the private sector. Over $7.7 million has been contributed by corporations, foundations, and individuals, including NCSSM parents, for programs, capital improvements, endowment, and student enrichment.

Elementary Schools

At the elementary level, special schools for the gifted are mostly private. An example is The Oaks Academy, an independent school for gifted children in grades K–6 that also offers a special preschool program for three- and four-year-olds. The Oaks Academy serves the northwest area of Houston, Texas, in Harris County.

Admission is competitive. Students are evaluated on the basis of academic and intellectual abilities as well as on qualities of character, leadership, and creative potential. Screening includes meetings with two psychologists who administer IQ, development, and achievement tests; interviews of prospective students and their parents; and questionnaires assessing prior school experiences.

The curriculum at The Oaks Academy is designed to provide for the development of strong basic skills, along with differentiated lessons appropriate for able learners. The program is highly individualized and flexible. There is heavy emphasis on building self-concept and a capacity for self-directed learning. Students are expected to strive for mastery in all content areas and to apply their knowledge in individual and group projects. Time is provided for the pursuit of special interests, for creative expression, and for the development of special talents in the fine arts. Violin, cello, piano, harp, painting, and creative dramatics are all offered on campus. Foreign language begins in preschool and is expanded throughout the grades. Computer literacy and programming are taught along with mathematics, science, social studies, reading, and language arts. In mathematics, the school teaches the Comprehensive School Mathematics Program (Heidema et al. 1978–86) described in the earlier discussion of elementary school programs.

The ratio of full-time teachers to students is one to fifteen. In addition, the school employs specialists in foreign language, art, music, physical development, and library science, along with an administrative staff, which reduces the ratio to one adult for every six children. Maximum class size is twenty in the upper grades. Eighty percent of the faculty hold advanced degrees, and all have had special training and many years of teaching experience.

A second elementary school example is found in the Seattle Country Day School, a special education program designed to meet the needs of intellectually gifted youngsters who also exhibit strong creative problem-solving abilities. The school was organized in 1963 as a nonprofit coeducational K–8 elementary school; it has full Washington State accreditation and enrolls 210 students. Admission is based on high academic ability, social adjustment, emotional stability, and previous academic achievement. Screening includes an individual IQ test administered by the school psychologist as well as an interview with parents and the child.

The school's basic objective is for students to develop their intellectual curiosity, creativity, and capacity for critical thinking. All students participate in an academic program that includes mathematics, science, language arts, social studies, French, computer literacy, art, music, drama, physical education, and health. Elective and enrichment courses are offered on a rotating basis from a curriculum that includes additional laboratory science, advanced mathematics, robotics, aerospace studies, creative writing, urban studies, art, dance, drama, poetry, public speaking, bridge, chess, and skiing. Numerous guest speakers and field trips are incorporated into both basic and elective curricula.

Class size averages sixteen. Primary-grade children remain with the homeroom teacher for the academic core subjects; middle school children change classrooms and teachers for each subject. Within each class students may be divided into smaller cluster groups. According to the school, the goals of this method are to place students into small classes with their mental peers and to provide skilled, imaginative, well-educated teachers possessing special training in gifted education.

Regional Centers

To augment the programs of comprehensive schools, regional centers offer courses for gifted and talented students that provide opportunities to engage in work beyond that offered in their own schools. In many cases the courses are taught in conjunction with a nearby university, as in Syracuse University's Project Advance (1985), which offers students college credit for courses completed in high school. Many other universities, such as the University of Maryland, Northwestern University, and the University of Southern Mississippi, offer special classes for gifted high school students during the academic year or during the summer months (BOCES I 1979).

The Mathematics Education for Gifted Secondary School Students (MEGSSS) project began as a program of the CEMREL educational laboratory in Saint Louis, Missouri. It initially enjoyed "National Model Project" status with the Office of Gifted and Talented. Now a not-for-profit private corporation, Project MEGSSS operates a center that qualified students from the Saint Louis metropolitan area can attend for their mathematics instruction. Students in the program participate in lieu of regularly scheduled mathematics classes at their home schools (Kaufman, Fitzgerald, and Harpel 1981).

The State University of New York at Buffalo's Gifted Math Program (GMP) was derived from the MEGSSS model and enjoys a strong and unique university-community bond. Approximately 250 students from about eighty-five schools participate. A central feature of the program is that each September a new group of seventh-grade students embarks on the full six-year school-college mathematics program at the university. Special agreements have been made between local schools, the state education department, and the university to permit students to obtain school, state regents, and university credit for their work in the program. Local public, parochial, and private schools view GMP as an extension of their own programs and as the university's commitment to the western New York community.

Nominations of sixth graders are solicited early each year from school administrators and parents. Families attend information meetings before screening is done. Students are selected on the basis of a battery of tests, and a private family conference is held to help insure that expectations and responsibilities are understood and that important channels of communication are established. Parents provide students with transportation to the university, where classes are scheduled on Mondays and Wednesdays from

September through June. Through communications with individual schools, students receive grades on their standard report cards, and supplemental reports are regularly issued to parents directly from GMP.

During the first four years of the program students study from the *Elements of Mathematics* (EM) books 0, 1, 2, 3, and Problem Book B (Martin 1970–1983). In addition, during the fourth year students study from *Precalculus Mathematics with a Computer* (Stover, Rising, and Schoaff 1985), and they are issued a pocket computer to assist them in the year's work. Although much of the curriculum of the first four years addresses secondary school content, college-level mathematics topics such as logic and abstract algebra are introduced in the first year and increased in proportion during subsequent years. This latter work can be converted, at the parents' option, to regular university credit. In the fifth and sixth years, students are enrolled in special sections of university courses in discrete mathematics and calculus. Those students study from *Discrete Algorithmic Mathematics* (Mauer and Ralston 1984) and *Calculus and Analytic Geometry* (Stein 1983). Students can accrue twenty-four university credits for the work they accomplish in the six years of the program. Further explanation of the curriculum and credit can be found in "The Gifted Math Program at SUNY at Buffalo" (Krist 1985).

The program, which has been in existence since 1980, recently graduated its first class, all of whom received scholarships and distinguished honors. They are currently studying a wide variety of curricula at prestigious universities throughout the country.

Funding for the program is provided by the university, private foundations, and fees paid by parents. Enough support is generated to allow for fees ($125 per semester in 1986) to be reduced or eliminated by parent request. Approximately 10 percent of the participating families choose this option. No such special provision is made for tuition for university credit, which in 1986 was $40 per credit hour.

The GMP teaching staff consists of both secondary teachers from local schools and university faculty. Each is selected on the basis of interest in the program, strong mathematics background, and reputation for excellence in teaching. Many are award winners, and all have at least master's degrees; 30 percent have doctorates. The program employs one part-time secretary.

To increase participation by minority students and females, the program has established the Gifted Math Program Saturday Enrichment Series (GMP/SES) for elementary students in the cities of Buffalo and Niagara Falls. With the aid of a state education grant, more than 250 fourth, fifth, and sixth graders attend a series of eight Saturday morning sessions at one of the two sites. The intent is to generate interest in mathematics, increase test-taking skills, and develop student ability to solve problems. Teachers and students in GMP/SES use the *Challenge: A Program for the Mathematically Talented* series (Haag et al. 1986), which was written by many of the same authors as the EM series. Parents and teachers routinely observe these and standard GMP classes, and many of them participate regularly in the class activities.

GMP's association with the university gives a serious academic tone to all its activities, and both the university and GMP benefit. Students have easy access to university facilities; they can use and withdraw books from the libraries and take advantage of athletic and cultural events on campus. Many have cultivated the informal mentorship of university faculty from various departments. Students see themselves as part of the university community, and many have chosen to pursue their college studies at the school. An annual symposium has exposed students to outstanding faculty from various departments and even to a Nobel Prize laureate. In a very friendly setting, students often see important people doing significant work.

As with any pull-out program, there are disadvantages as well as advantages. Students do have an opportunity to work with their peers, but they also can become isolated from other students in their home schools. The program has no formal counseling, and very often the time needed to cultivate fragile relationships with young people is lacking. Parents and students are given the home telephone numbers of the entire staff and are encouraged to call them directly to ask questions or receive help. Although many students do call, most do not; and the daily contact of student and teacher normally found in school programs is limited. However, parents, students, and staff strive to maintain direct and open communication and to take maximum advantage of the time they have together.

A university advisory committee assists the program directors. It is co-chaired by the associate deans of the Faculty of Educational Studies and the Faculty of Natural Science and Mathematics; its membership includes the chairs of the Departments of Learning and Instruction, Mathematics, and Computer Science; the GMP directors; a faculty member from outside any of the named departments; and an administrative liaison from the vice-provost's Office of Undergraduate Education. Another advisory committee is composed of the program directors and local school mathematics supervisors. Parents have an informal committee, and good communications are maintained between program personnel and the state education officials who routinely evaluate the program.

In-School Alternatives for Elementary Pupils

One approach to providing for mathematically gifted elementary school children is through mathematics specialists who serve all the schools in a district. A model of this approach is found in the Oceanside Union Free School District, Oceanside, New York, where one mathematics specialist provides enrichment in a variety of ways.

First, during the year the specialist provides one four-day series of mathematics enrichment sessions in every classroom, grades one through five, in each of the district's five elementary schools. These lessons involve students in hands-on problem-solving experiences with different materials at each grade level.

Next, the specialist works with students within the districtwide gifted program. Approximately 10 to 15 percent of fourth- through sixth-grade children are involved in this pull-out program, which includes all subject areas and is taught by two teachers of the gifted. One day each week these students are bused to a selected school where they participate in this program from 9:30 a.m. until 1:30 p.m. Currently, the elementary mathematics specialist is working with one group of fourth graders for one hour a week during that time period, involving them in advanced mathematics and problem-solving experiences.

In addition, elementary school pupils in the Oceanside district have the opportunity for enrichment through the before-school Mathematics Olympiad training sessions organized by the mathematics specialist. These sessions, held one day a week at each of the five schools, begin at 7:40 a.m., and large numbers of fourth- through sixth-grade pupils attend regularly. The specialist works with one school and has trained others to work in the other schools. The specialist also attempts to provide some opportunities for third graders to participate by working one morning a week with a third-grade group at one school and another morning a week with a combined third- and fourth-grade group at another school. As a result of these activities, there has been a significant increase in the number of students participating in similar mathematics contests at the junior high and high school levels. Recently, four out of five students participating on the award-winning high school mathematics team were former elementary school "mathletes."

The Oceanside district also operates one period during the school day designated as the "X Period," during which special events and programs can be scheduled. The elementary mathematics specialist uses this period in a variety of ways as well. For example, the specialist works in each of the elementary schools once a week with a special group of second graders who have been identified as "especially strong thinkers" in mathematics. She uses materials she developed that are similar to the Mathematical Olympiad training materials used for the upper grades (Lenchner 1983). The specialist also uses the "X Period" to work with three outstanding students (one second, one third, and one fourth grader) in each of three schools on a one-to-one basis. The teacher, principal, and mathematics specialist have identified these students as truly gifted in mathematics and in need of more than what is offered in whole-class or small-group programs.

A second approach to providing mathematics instruction for gifted elementary pupils is to group them in a special class. This is the model adopted in the "Windows" program, a special magnet program operated by the Packanack School in Wayne, New Jersey. Twenty-five selected fourth and fifth graders in the public school district are bused to this school, where they are assigned to a special class taught by an elementary school teacher with advanced training in mathematics and mathematics education. Selection for this program is based on a combination of criteria that includes IQ scores, reading and mathematics achievement test scores, and recommendations.

The mathematics curriculum adopted for the Windows program is interdisciplinary in focus, and mathematics is taught through special units of instruction in science, social studies, and language arts. An example is the unit "Mathematics in Nature: Patterns, Structure and Functions," which introduces students to transformational geometry. A wide variety of concrete materials are used for problem solving and the development of logical reasoning. The textbooks used are one grade level higher than the students' grade placements, and supplementary materials, including books intended for middle school, also are used. Students use the four computers in the classroom for simulations, problem solving, and Logo and BASIC activities.

A third example of the ways in which elementary schools can provide for gifted pupils is found in the "Newton Advanced Challenge Program" in Newton, Massachusetts. In this district, all fifteen elementary schools use the *Developing Mathematical Processes* (DMP) program (1975), which was implemented over several years through the provision of elementary mathematics specialists to introduce classroom teachers to the program and to assist them with implementation of the curriculum. Although the district reported that they found DMP to be a "very rich" program for their students, nonetheless differences in mathematical abilities became pronounced by the fourth grade. This prompted development of additional mathematical opportunities through the Newton Advanced Challenge Program (NCP).

The NCP has two components: a challenge group component and an advanced learner component. The challenge group consists of approximately 15 percent of the children in grades 4 through 6, who are identified through a combination of teacher and parent recommendations, achievement test scores, and self-selection. Lateral enrichment in mathematics takes place within the classroom through the efforts of the classroom teachers and a staff of four specialists who provide assistance to the teachers, primarily in the areas of mathematics and science. This assistance includes helping to teach the classes, identifying appropriate resource books and materials, and developing guidelines for selecting enrichment activities.

The advanced learner component is primarily for fourth- through sixth-grade students; only 1 to 2 percent of the pupils (about fourteen to twenty students districtwide) participate. Twice weekly they meet in their own schools with a mathematics specialist. In these sessions fourth and fifth graders explore extensions of the topics taught at their grade levels as well as more advanced topics. Sixth graders cover the mathematics typically included in the sixth- and seventh-grade curricula along with several new topics, the use of computers, and BASIC. The specialists also serve as mentors for the students, facilitating independent investigations into a variety of mathematical topics including probability, statistics, spatial activities, exploring number patterns and sequences, and number theory. The students who participate in the advanced learner program are excused from their regular classroom placements; they are not held responsible for completing all the ordinary grade-level mathematics activities.

Advanced Curricula for Secondary Schools

A program that has gained widespread acceptance in secondary schools is the Advanced Placement (AP) program. This is a nationwide program that allows high-ability students to complete college-level work while still in high school. Sponsored by the College Entrance Examination Board, an independent association of schools and colleges, the AP program includes courses in a variety of subjects, including calculus, physics, English, and computer science, leading to special AP examinations that can qualify a student to receive college credit from a large number of colleges and universities.

Scores on the AP examinations are summarized on a five-point scale: (1) no recommendation, (2) possibly qualified, (3) qualified, (4) well qualified, and (5) extremely well qualified. However, although the AP tests are scored uniformly, colleges and universities are not uniform in their acceptance of AP credits. Further, not all schools offer AP courses; this is especially true among small schools.

A somewhat less familiar alternative is the International Baccalaureate (IB) program, a curriculum for serious senior high school students that is featured in a growing number of schools around the world. The full IB program is a rigorous one that provides the students with a well-rounded educational experience that, in the United States, often qualifies graduates for admission to prestigious universities with sophomore standing. To earn the IB diploma, the student must pass examinations in six areas:

1. Language A—the language of instruction in the student's school
2. Language B—a second language
3. Study of man—one of the following: history, geography, economics, philosophy, psychology, social anthropology, or business studies
4. Experimental studies—one of the following: biology, chemistry, physics, physical science, or scientific studies
5. Mathematics
6. One of the following additional options: art, music, a classical language, a second Language B, an additional choice from categories three through five above, computer studies, or a special syllabus developed by IB schools

Of the six courses, three must be studied—and their examinations passed—at a higher level, and the other three at a subsidiary level. In mathematics the student may choose to take an examination from one of several levels depending on how much calculus the student has studied during high school. The candidate for the full diploma must also take Theory of Knowledge, a specially designed interdisciplinary course on the philosophy of learning that requires the student to reflect on the nature of learning, the nature of his or her own learning, and the relationships among different modes of learning. In addition, IB candidates are required to participate in a minimum number of hours devoted to creative, aesthetic, or social service activities. Finally, the candidate is required to produce an extended essay based on independent research.

A student who does not meet all the requirements above may still qualify for a certificate in certain courses. Many colleges and universities treat an IB certificate in a given course as equivalent to passing an AP examination. Students who pass an IB examination at a specified level (usually four points on a seven-point scale) are granted credit for the course. A principal difference between the IB program and the AP lies in the fact that the IB is a comprehensive curriculum with an international emphasis focused on problem solving and the interrelation of knowledge, whereas the AP exists in the form of separate courses (Cox, Daniel, and Boston 1985).

Summer Programs

Perhaps the most notable common denominator of summer programs for the gifted is warm weather. In other respects these programs are as diverse as any within the school. They may be day programs or residential; elementary, middle, or high school level in content; broad in scope or focused specifically on mathematics; operated by local schools, consortia of school districts, universities, or even private enterprises.

Representative of numerous local summer programs is the Gateways Summer School for the gifted and talented in San Diego, California. Courses are offered for students in grades 1 through 12, but the bulk of the enrollment is composed of elementary and middle school pupils. The main goal of the program is to provide valid learning experiences with significant content in an appealing setting. Classes are offered in a wide spectrum of subjects that include electricity, biology, rocketry, photography, oil painting, short story writing, and magic. Mathematics offerings include the study of tessellations and Escher-like designs, the construction of topological curiosities, and the creation of mathematically generated art of all types. Many courses are offered at two levels to meet the different needs of elementary and secondary students.

The summer institute for mathematically gifted high school students held at the Ohio State University provides an eight-week residential program for forty ninth- through twelfth-grade students selected from across the United States. These participants work on problems in number theory, combinatorics, algorithmic mathematics, and computer science. The objective is to focus on thinking, intellectual challenge, and the development of attitudes that foster productive thinking, and students are given unique opportunities

4. Four long segments are of equal length and four short segments are of equal length. The long segments are twice the length of the short ones. Use all eight segments and make exactly three congruent squares.

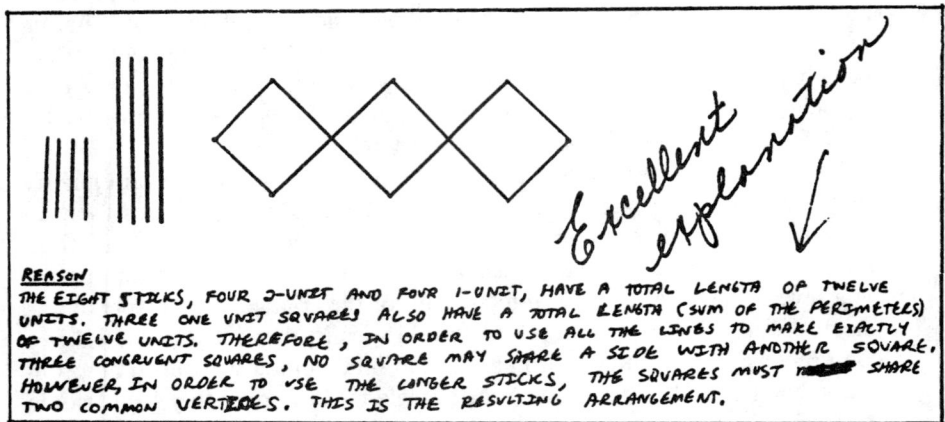

Excellent explanation ✓

REASON
THE EIGHT STICKS, FOUR 2-UNIT AND FOUR 1-UNIT, HAVE A TOTAL LENGTH OF TWELVE UNITS. THREE ONE UNIT SQUARES ALSO HAVE A TOTAL LENGTH (SUM OF THE PERIMETERS) OF TWELVE UNITS. THEREFORE, IN ORDER TO USE ALL THE LINES TO MAKE EXACTLY THREE CONGRUENT SQUARES, NO SQUARE MAY SHARE A SIDE WITH ANOTHER SQUARE. HOWEVER, IN ORDER TO USE THE LONGER STICKS, THE SQUARES MUST ~~not~~ SHARE TWO COMMON VERTICES. THIS IS THE RESULTING ARRANGEMENT.

3. Find the smallest number, N, such that:

N/10 leaves a remainder of 9;   N/5 leaves a remainder of 4;
N/9 leaves a remainder of 8;   N/4 leaves a remainder of 3;
N/8 leaves a remainder of 7:   N/3 leaves a remainder of 2;
N/7 leaves a remainder of 6;   N/2 leaves a remainder of 1.
N/6 leaves a remainder of 5;

Knowing that the value of N divided by 10 yields a remainder of 9 shows us that the last digit of N has to be 9. N divided by 9 yielding a remainder of 8 told us that N must be one less than a multiple of 90. It then dawned on us that the <u>Least Common Multiple of the numbers 10 through 2 minus 1 would give a value for N which would fit the specification</u>. The Least Common Multiple turns out to be $2^3 \cdot 3^2 \cdot 5 \cdot 7$ or 2520. Subtracting one from that gives us one value for N. 2519

→ *Good for you! It is "messy" to look for N directly — a much easier related problem is to find (N+1) first —*

to develop powers of observation, to experiment, to discover significant relations among the objects of experimentation, to use counterexamples and proof, and to develop precise and concise language.

The Ohio program has been in existence for over twenty years (although not always at the present site), and students often return to it not only during their subsequent years of high school but also during their undergraduate and graduate studies. Admission to the program uses a unique approach in which students are asked to solve problems meant to test their ingenuity and interest rather than the breadth of their experiences. It is suggested to the applicants that they spend several weeks on the problems, and they are encouraged to explore the topics through reading if a suitable library is available. The exact topics of study vary slightly from year to year, but the aim remains constant: to provide students with opportunities to acquire experience in scientific thinking while remaining within the field of mathematics (Ross 1978).

Another summer program with a specific mathematics curriculum is the Summer Institute for the Arts and Sciences held at Northern Michigan University in Marquette. This is one of a larger set of such programs for the gifted operated since 1982 under the auspices of the Michigan State Board of Education. Each is held on a university campus and combines both a science-oriented and an art-oriented curriculum. The Northern Michigan program draws 125 entering eleventh and twelfth graders for a two-week residential program in mathematics and the visual arts; approximately two-thirds of the group are mathematics students.

Announcements of the institute are sent to the local schools, which are asked to nominate qualified students. Nominations are forwarded to the Intermediate School District, a unit ordinarily composed of school districts from several counties, where they are screened by a selection committee. Criteria include high school academic records, teachers' recommendations, and students' personal statements of goals. Recommendations of the committee are then forwarded to the Michigan Department of Education, where final selections are made. Students in the institute pay their own transportation costs and a $150 fee; all other costs for tuition, room, and board are paid by the state. In many cases, local schools or civic organizations offer scholarships to cover all or part of the fee.

The primary focus of the mathematics curriculum is on problem solving, and all students take a two-week course that helps them to develop and apply problem-solving skills to a range of mathematical topics. For this part of the program students are assigned in teams of five, and they work cooperatively to solve problems and prepare written accounts of their solutions. Stress is placed on communication, and written solutions are evaluated on the basis of completeness, accuracy, clarity, organization, and originality. Additional problems are posed as individual challenges to supplement the group work.

Each student also takes two one-week elective courses that treat in greater depth topics not usually studied in high school. Choices include probability, limits, proof and provability, algorithm design, and number theory. Shorter elective sessions (ordinarily two to four hours each) introduce students to a range of topics and new branches of mathematics including Logo, word processing, group theory, topology, the mathematics of economics and finance, recreational mathematics, and more.

Some time each week is also devoted to interdisciplinary experiences in which both mathematics and art students participate. Offerings in the interdisciplinary program change each year, but they have included such topics as drawing Escher-like tessellations, mathematics and music, computer graphics, and paper folding. Other sessions offer participants the opportunity to discuss the mathematical aspects of various careers with representatives from engineering, medicine, law, education, and other areas. An open forum called "Straight Talk on Being Smart" allows students to discuss candidly their feelings, concerns, questions, apprehensions, and opinions about what it means to be gifted and talented. A full program of physical education and recreational activities completes the curriculum.

Mathematics courses during the institute are taught by regular and visiting university professors with special interest in gifted education and by high school teachers with expertise in the field. The schedule includes large- and small-group presentations and laboratories; the instructional emphasis is on understanding and proof. In addition, a number of high school teachers serve as teacher-mentors who work with the students in the problem-solving labs, make presentations during the short subject offerings, and serve as academic and personal counselors to the students. Faculty members live in the dormitory with the students, which contributes to an open, informal environment.

Students receive neither grades nor credit for their work, but a written report is sent to the home school mathematics teacher of each individual. These reports, prepared by the teacher-mentors and faculty, provide the schools with specific information about the student's achievements and study habits. Feedback from home school teachers indicates that they appreciate these reports and find them helpful both in providing for the students after the program and in identifying candidates for future programs.

In evaluating the program, students repeatedly indicate that the institute helps them to change their image of mathematics as a collection of rules and algorithms and to realize that it is a complex, multifaceted discipline whose patterns and regularities make it predictable and understandable, that ideas that at first appear unrelated often come together in surprising ways, and that there are depths of mathematics previously undreamed-of by them. They also report that during the institute they develop new insights into, and appreciation of, their own abilities and interests in mathematics, learn to discuss and defend their ideas with others, come to a more realistic assessment of their own talents with respect to other gifted students, and recognize that when it comes to mathematics, "I have a lot to learn."

Magnet Programs

Magnet programs are designed to attract students of high ability and interest in a particular field of study. An example is the "Summatech" magnet for mathematics, science, and technology operated by the Minneapolis, Minnesota, Public Schools. This four-year program is housed in one of the city's high schools but is open to interested and qualified students from anywhere in the district. The goal of Summatech is to acquaint students with "what it takes to make it in a world dependent upon technology" (Minneapolis Public Schools Program Guide).

Summatech is not a program only for the gifted—it was designed to attract students seeking excellent preparation for higher education and a strong background for technology-related fields. Applications are sought from students who have an aptitude in science and mathematics and who "are intellectually curious and inventive." Students are encouraged to apply for admission in time to enter as ninth graders, and once enrolled, they are expected to complete four years in the program. Selection criteria include, in addition to academic records, recommendations from teachers and counselors and a statement from the student describing his or her interest in the program. A minimum grade point average of 2.5 overall and 2.3 in Summatech classes is required for continuation. Transportation is provided by the district.

Summatech students receive their mathematics, science, and technology instruction in a three-hour block of time that allows for greater interdisciplinary emphasis in the curriculum. Other activities include extended field trips, participation in competitions, individual and cooperative projects, mentorships, and internships. Professionals from industry and from higher education provide guest lectures and discussions, and they work with the Summatech students in areas of special interest. In addition, students must take three other required or elective courses selected from the curriculum of the larger school, and they are given the option for a seventh period in order to take additional electives such as music or a foreign language.

The mathematics curriculum includes elementary and intermediate algebra, geometry, trigonometry, mathematical analysis, and Advanced Placement calculus. The standard content of these courses is augmented with additional topics such as linear programming, transformational and non-Euclidean geometries, topology, probability, statistics, mathematical structures, linear algebra, deductive logic, and Boolean algebra. The science component is organized in an interdisciplinary manner with a thematic focus—for example, "energy" for each of the first three years. In the context of the chosen focus, students study physical, chemical, and biological properties and their interactions. The fourth year of the science program is one of enrichment, with the opportunity to select classes and independent research according to individual interests.

A second type of magnet program can be found in the Center for the Arts and Sciences (CAS) operated by the School District of the City of Saginaw, Michigan. This magnet school for gifted students in grades 7 through 12 offers programs in mathematics and science, global studies, language arts, dance, instrumental music, theater, visual arts, and voice and keyboard. The school is open to all students in the district; selection is based on School and College Ability Test (SCAT) scores, transcripts, letters of recommendation, and interviews. In addition, music, theater, and dance students must audition; art students must submit a portfolio.

Students who are accepted for one of these programs are released from their home schools to attend CAS on a half-day basis. There is no tuition to the students, and the school provides transportation for city residents. Students attend CAS either from 9:00 to 11:30 a.m. or from 12:45 to 3:15 p.m. daily. Each enrolls for only one of the courses named above, and the entire time is spent in that discipline. Mathematics and science form an integrated program, and students divide the two and a half hours between the two. They also take courses in English, social studies, physical education, and various electives during the half day spent in their home schools.

The center enrolls approximately 375 students; nearly 100 of these are enrolled in the mathematics and science program. There are three faculty members for the mathematics and science courses. Although students' programs vary according to their entry-level placement, a typical program for a student enrolled in mathematics and science might include algebra, earth science, and computer programming (grade 7); geometry and physical science (grade 8); intermediate algebra, trigonometry, and biology (grade 9); precalculus mathematics and chemistry (grade 10); calculus and physics (grade 11); and additional individualized options (grade 12).

Mentor Programs

Ysleta Girls Count is a summer enrichment and mentor program designed to introduce junior high school girls to opportunities in mathematics and mathematics-related careers. The program is operated by the Ysleta Independent School District in El Paso, Texas.

Forty students who have completed seventh-grade honors mathematics courses and who have scored at the 97th percentile or above on sixth-grade standardized tests are recruited from throughout the district and brought together for a week-long summer program that includes class work, field trips, and the chance to interact with female role models during luncheons, panel discussions, and other contacts. The daily schedule is a demanding one that includes morning classes focused on creative concepts in mathematics. Afternoon tours of a local hospital and medical school, the White Sands Missile Range, the University of Texas at El Paso, and a local bank are included in the activities. During each of the tours the students are introduced to women holding high-level positions in the facility.

During the program, groups of girls are matched with a female mentor, most of whom also participate in the summer program activities. The women who serve as mentors hold responsible positions in engineering, computer programming, research, administration, and a variety of other careers. Each mentor is expected to contact her students at least monthly during the following school year, either visiting the girls in person or by telephone. During the year each group also prepares a project that is presented and critiqued in the spring.

A two-day follow-up workshop is held during the summer following the initial program, with activities similar to the first summer. Additional contact between the program directors and the participants during the remainder of the girls' high school years assures the participants of the continued interest of the program sponsors.

Mathematics Contests

Mathematics contests offer unique and important opportunities to extend the education of gifted students. By participating in the various kinds of competitions available, a talented student can develop problem-solving skills at a higher level than is required in a typical mathematics curriculum, explore topics not usually covered in school, and interact with other mathematically talented individuals.

An increasing number of different kinds of contests exist at the local, regional, and national levels. Most are intended for high school students, but recently similar competitions have been organized at the junior high and elementary school levels. Participation in these events has become a regular extension of the mathematics program for many schools.

There are two basic types of mathematics competitions: problem-solving contests and contests requiring the development of a project. In the former, the student is usually required to solve a series of challenging problems within a given time frame (usually one or two hours) without the use of reference materials. These contests can have either an objective format, usually multiple choice, or an essay format requiring open-ended responses.

Among the most well known annual mathematics contests for secondary school students in North America are those administered by a committee of the Mathematical Association of America. These include the American High School Mathematics Examination (AHSME), the American Junior High School Examination (AJHSME), the American Invitational Mathematics Examination (AIME), and the USA Mathematics Olympiad (USAMO). Each of these competitions serves a specific purpose.

The oldest of the contests is the AHSME, often referred to as the MAA contest. Its main purpose is "to spur interest in mathematics and develop talent through the excitement of friendly competition at solving intriguing problems in a timed format" (Mauer and Mientka 1982). The examination contains items with a range of difficulty and is intended for average students who enjoy mathematics as well as for the most exceptional mathematics students. The thirty multiple choice items that make up the ninety-minute examination can be solved using the content of the precalculus mathematics program.

Another purpose of the AHSME is to identify and give recognition to students with exceptional talent in mathematics. It serves as a first step that can lead to eventual participation in the International Mathematical Olympiad, considered the most difficult and prestigious secondary level mathematics examination in the world. A student must earn a score of 100 out of a possible 150 on the AHSME in order to qualify for the next level of competition, the AIME.

The AIME was first offered in 1983 as an intermediate examination between the AHSME and the USAMO. Approximately 1 percent of the 360 000 students who took the AHSME during 1985–86 qualified to take the AIME. In its present form the AIME contains fifteen problems, to be completed in two and a half hours. Again, the problems can be solved with precalculus techniques, but an essay format is used. Students' answers are graded either right or wrong; there is no partial credit.

Approximately fifty of the top scorers on the AIME in North America are invited to the USAMO. This third level of competition provides still further challenge, intended to identify the best candidate to represent the United States in the International Mathematical Olympiad (IMO). The top eight scorers on the USAMO are named "winners" and invited to the United States training sessions for the IMO. In addition, USAMO scores are the basis for extending invitations to approximately twenty-four more students, who also attend the training sessions; all but the top ten students invited to those sessions must be nonseniors. The team is chosen from among the top ten; the other participants train for future competitions. Similar olympiad competitions and training sessions take place in Canada, which has its own IMO team.

The newest competition in this sequence of contests is the AJHSME, which was begun in 1985. This competition, intended for seventh- and eighth-grade students but open to any student who has not completed eighth grade, consists of twenty-five multiple choice items to be completed in forty minutes. The exam questions include mathematics topics from typical seventh- and eighth-grade programs presented in the form of challenging problems.

Many other contests are organized as league competitions with teams representing their schools at various levels of competition. An example is the Atlantic Pacific Mathematics League, a national competition intended for students in grades 9 through 12. Using a short-answer problem-solving format, six contests are held throughout the year. Over ten thousand students typically participate in this league annually.

Similarly, the American Regions Mathematics League primarily involves tenth- through twelfth-grade students from many different states. Teams of fifteen students participate in an annual meet that includes both short-answer questions and power questions to be answered in essay form. Some questions are intended to be answered by individual students, others by teams, and still others in a relay format. The competitions have been held at Pennsylvania State University and have attracted between 750 and 800 students and between 100 and 150 teachers.

A recently organized national mathematics league competition for seventh- and eighth-grade students is MathCounts. Schools are provided with study materials that include problems supporting and extending the seventh- and eighth-grade curriculum. School teams compete on a local, regional, state, and national level. Certificates of participation, trophies, and medallions are awarded at the different levels of competition; state winners earn a trip to Washington, D.C., for the national competition.

The Mathematical Olympiads for Elementary Schools (MOES) is an example of a successful team competition for younger pupils. Teams of up to thirty-five students participate in five contests held at monthly intervals during the school year. Each contest consists of five challenging problems, each with a time limit. Although the majority of participants are fifth and sixth graders, capable fourth-, third-, and even second-grade pupils have participated successfully.

Individual participants in the MOES receive a special certificate; other awards and trophies are given to reward individuals who achieve above a specified minimum score, for the best individual performance on a team, and for the highest team scores. Sample materials and in-service courses have been developed to assist teachers in preparing students for this competition.

Participation in the problem-solving contests available at all grade levels can provide gifted students with opportunities to solve problems that are more challenging than those they encounter in their mathematics classes and to compete with other students who are interested and talented in the subject. Often this activity is supported by a mathematics club or team, with a teacher serving as advisor or coach.

The availability of a teacher-advisor to help students prepare for competitions is an important ingredient for success. Many of the benefits of participating in these competitions come from the preexamination and postexamination activities. The advisor also motivates students for the competition, helps them locate materials needed to prepare for the contests, and nurtures the development of specific problem-solving strategies as well as general test-taking skills.

Another form of mathematics competition is the mathematics fair. Here each student develops a project based on a mathematical topic or principle and prepares an exhibit to describe the project. Competitions take place at local, national, and international levels with prizes varying from certificates to cash awards and scholarships.

Many gifted students find the development of a project to be a challenging and exciting venture. Unlike a problem-solving contest, fairs do not operate under a rigid time frame, and students find more opportunities to exercise their personal interests and creativity. The projects they pursue require independent work and task commitment, characteristics that typify many gifted students.

The Westinghouse Science Talent Search is an annual competition that awards scholarships to graduating high school seniors for independent research projects that present evidence of research ability in science or engineering. The entrants' reports are each about one thousand words in length and provide an opportunity for students both to prove that they can plan and carry to completion some problem or project and to show that they have the ability to approach a problem with the creativity and originality that are essential for research. Historically, few awards have been made for work in mathematics, perhaps because potential participants fail to realize that there is a significant difference between showing originality and being original. A reading of the titles of award-winning projects from the past indicates that students' work need not be on the frontiers of knowledge in order to embody the quality of originality.

Useful resources for helping students choose topics for independent projects include the *Mathematics Project Handbook* (Hess 1982) and *Student Merit Awards—Middle School* and *Student Merit Awards—High School* (Sachs 1984). For a more complete listing of mathematics competitions, see *Mathematics Contests—A Handbook for Mathematics Educators* (Johnson and Margenau 1982).

4
Resources for Teachers of the Gifted

FOLLOWING are sources of further information on teaching the gifted. The first section lists names and addresses for the programs described in Chapter Three. The second section lists organizations and offices with a focus on gifted education.

Contacts for Programs Described in Chapter 3

Boston Latin School
78 Avenue Louis Pasteur
Boston, Massachusetts 02115

Bronx High School of Science
75 West 205 Street
New York, New York 10468

The Oaks Academy
16720 Stuebner Airline
Box 240
Spring, Texas 77377

Seattle Country Day School
2619 Fourth Avenue North
Seattle, Washington 98109

Gifted Mathematics Program
560 Baldy Hall
SUNY at Buffalo
Buffalo, New York 14260

Ms. Sandy R. Cohen
Oceanside Union Free School District
Oceanside, New York 11572

The North Carolina School of Science and Mathematics
1912 West Club Boulevard
Durham, North Carolina 27705

Dr. Gloria Sanok
Windows Program
Packanack School
Wayne, New Jersey 07470

Dr. William Radomski
Newton Advanced Challenge Program
100 Walnut Street
Newton, Massachusetts 02160

International Baccalaureate North America
200 Madison Avenue
New York, New York 10016

Ms. Carolyn Wood
Gateways Summer School Program
3818 Mott Street
San Diego, California 92122

Professor John Kiltinen
Professor Donald Zalewski
Department of Mathematics
Northern Michigan University
Marquette, Michigan 49855

Summatech
North Community High School
1500 James Avenue North
Minneapolis, Minnesota 55411

Center for the Arts and Sciences
115 West Genesee
Saginaw, Michigan 48602

Ms. Evelyn D. Bell, Assistant
   Superintendent
Ysleta Independent School District
9600 Sims Drive
El Paso, Texas 79925

(For AHSME, AIME, USAMO, and AJHSME contests)
Dr. Walter E. Mientka
American Mathematics Competitions
Department of Mathematics and
   Statistics
University of Nebraska
Lincoln, Nebraska 68588

Mr. David Rosen
Atlantic Pacific Mathematics League
P.O. Box 11242
Elkins Park, Pennsylvania 19117

Advanced Placement Program
College Entrance Examination Board
Princeton, New Jersey 08540

Professor Arnold E. Ross
Department of Mathematics
Ohio State University
231 West 18th Avenue
Columbus, Ohio 43210

Dr. Alfred Kalfus
American Regions Mathematics
   League
Department of Mathematics
Hofstra University, South Hall
Hempstead, New York 11550

MathCounts
National Society of Professional
   Engineers Information Center
1420 King Street
Alexandria, Virginia 22314

Mr. George Lenchner
Mathematical Olympiads for
   Elementary Schools
Forest Road School
Valley Stream, New York 11582

Westinghouse Science Talent Search
Science Service
1719 North Street N.W.
Washington, D.C. 20036

Organizations and Agencies

National Leadership Training Institute
   on the Gifted and Talented
c/o Ventura County Superintendent of
   Schools
535 East Main Street
Ventura, California 93009

World Council for the Gifted and
   Talented, Inc.
Box 218
Teachers College
Columbia University
New York, New York 10027

The Council for Exceptional Children
1920 Association Drive
Reston, Virginia 22091

School Science and Mathematics
   Association
126 Life Science Building
Bowling Green State University
Bowling Green, Ohio 43403

National Association for Gifted
   Children
4175 Lovell Drive, No. 140
Circle Pines, Minnesota 55014

National Council of Teachers of
   Mathematics
1906 Association Drive
Reston, Virginia 22091

REFERENCES

Abelson, Harold, and Andrea diSessa. *Turtle Geometry*. Cambridge: MIT Press, 1980.

Action for Excellence. Denver: Education Commission of the States, 1983.

Aiken, Louis. "Ability and Creativity in Mathematics." *Review of Educational Research* 43 (Fall 1973): 405–32.

Alexander, P., and J. Muia. *Gifted Education: A Comprehensive Roadmap*. Rockville, Md.: Aspen Systems Corp., 1982.

Alvino, J., R. McDonnell, and S. Richert. "National Survey of Identification Practices in Gifted and Talented Education." *Exceptional Children* 48 (1981): 124–32.

Bagley, R., Kenneth Frazee, Jean Hosey, James Kononen, Robert Siewert, Jan Speciale, and Doris Woodfield. *Identifying the Talented and Gifted*. Salem, Oreg.: Oregon Department of Education, 1979.

Barbe, Walter B., and Joseph S. Renzulli, eds. *Psychology and Education of the Gifted*. 3d ed. New York: Irvington Publishers, 1981.

Bishop, William. "Successful Teachers of the Gifted." *Exceptional Children* 34 (January 1968): 317–25.

Bloom, Benjamin S., ed. *Developing Talent in Young People*. New York: Ballantine, 1985.

BOCES I Regional Resource Center for Gifted and Talented Education. *Programs for the Gifted and Talented*. Buffalo, N.Y.: BOCES I, 1979.

Boston, Bruce O. "Starting a Gifted Program." In *Developing Elementary and Secondary School Programs*, edited by Bruce O. Boston. Reston, Va.: Council for Exceptional Children, 1975.

Boyer, Ernest L. *High School: A Report on Secondary Education in America*. New York: Harper & Row, 1983.

Brown, Stephen I., and Marion I. Walter. *The Art of Problem Posing*. Philadelphia: Franklin Institute Press, 1983.

Bushaw, Donald, Max Bell, Henry O. Pollak, Maynard Thompson, and Zalman Usiskin. *A Sourcebook of Applications in School Mathematics*. Reston, Va.: National Council of Teachers of Mathematics, 1980.

Clark, Barbara. *Growing Up Gifted*. Columbus, Ohio: Charles E. Merrill Publishing Co., 1983.

Cohn, Sanford J. "What Is Giftedness? A Multidimensional Approach." In *Gifted Children: Challenging Their Potential*, edited by A. H. Kramer. New York: Trillium Press, 1981.

Cole, Donald B., and Robert H. Cornell, eds. *Respecting the Pupil: Essays on Teaching Able Students*. Exeter, N.H.: Phillips Exeter Academy, 1981.

Coleman, Laurence J. *Schooling the Gifted*. Menlo Park, Calif.: Addison-Wesley Publishing Co., 1985.

Council for Exceptional Children. "Digests on the Gifted." Reston, Va.: The Council, 1985.

Cox, June, Neil Daniel, and Bruce O. Boston. *Educating Able Learners*. Austin, Tex.: University of Texas Press, 1985.

David, Edward E., Jr. "The Federal Support of Mathematics." *Scientific American*, May 1985, pp. 45–51.

Davis, Steve, and Phyllis Frothingham. "A Special School in North Carolina." In *The Secondary School Mathematics Curriculum*, edited by Christian R. Hirsch, pp. 184–88. Reston, Va.: National Council of Teachers of Mathematics, 1985.

Developing Mathematical Processes (DMP). Nashua, N.H.: Delta Education, 1975.

Dugdale, Sharon, and David Kibbey. *Green Globs*. Pleasantville, N.Y.: Sunburst Communications, 1986.

Engel, Arthur, Bert Kaufman, and Edward Martin. *Elements of Mathematics Book B Problem Book*. St. Louis: CEMREL, 1975.

Fehr, Howard, et al. *Unified Modern Mathematics*. Secondary School Mathematics Curriculum Improvement Study. New York: Columbia University, Teachers College Press, 1968–1972.

Feldhusen, John F., J. W. Asher, and S. M. Hoover. "Problems in the Identification of Giftedness, Talent or Ability." *Gifted Child Quarterly* 28 (Fall 1984): 149–51.

Galbraith, Judy. *The Gifted Kids Survival Guide*. Minneapolis: Weatherall, 1983.

―――. *The Gifted Kids Survival Guide for Ages 10 and Under*. Minneapolis: Free Spirit, 1984.

Gallagher, James J. "Issues in Education for the Gifted." In *The Gifted and the Talented*, edited by A. H. Passow. Chicago: National Society for the Study of Education, 1979.

―――. *Teaching the Gifted Child*. 2d ed. Boston: Allyn & Bacon, 1975; 3d ed., 1985.

Gardner, Martin. *The Ambidextrous Universe*. New York: Charles Scribner's Sons, 1979.

Gensley, J. "Parent Perspective: Only the Learner Can Learn." *Gifted Child Quarterly* 13 (1969): 49–50.

Getzels, J., and J. Dillon. "The Nature of Giftedness and the Education of the Gifted." In *Second Handbook of Research on Teaching*, edited by Kenneth Travers. Washington, D.C.: American Educational Research Association, 1973.

Getzels, Jacob W., and Philip W. Jackson. *Creativity and Intelligence*. New York: John Wiley & Sons, 1962.

―――. "The Meaning of 'Giftedness': An Examination of an Expanding Concept." *Phi Delta Kappan* 40 (1958): 75–77.

Gilberg, Jody A. "Formative Evaluation of Gifted and Talented Programs." *Roeper Review* 6 (September 1983): 43–44.

Gowan, John C., Joe Khatena, and E. Paul Torrance, eds. *Educating the Ablest*. Itasca, Ill.: F. E. Peacock, 1979.

Guilford, Joy P. "Creativity." *American Psychologist* 5 (1950): 444–54.

―――. *The Nature of Human Intelligence*. New York: McGraw-Hill Book Co., 1967.

―――. "Three Faces of Intellect." *American Psychologist* 14 (1959): 469–79.

Haag, Vincent, Burt Kaufman, Edward Martin, and Gerald Rising. *Challenge: A Program for the Mathematically Talented*. Menlo Park, Calif.: Addison-Wesley, 1986.

Hagen, E. *Identification of the Gifted*. New York: Columbia University, Teachers College Press, 1980.

Heid, M. Kathleen. "Characteristics and Special Needs of the Gifted Student in Mathematics." *Mathematics Teacher* 76 (April 1983): 221–26.

Heidema, Clare, et al. *Comprehensive School Mathematics Program (CSMP)*. Kansas City, Mo.: Midcontinent Regional Educational Laboratory, 1978–86. (Distributed by Sopris West, Inc., Longmont, Colo.)

Henig, Robin Marantz. "Smart Kids, Hard Questions: The Challenge of the Gifted Child." *Washington Post Health*, 2 September 1986, pp. 13–16.

Hess, Adrien L. *The Mathematics Project Handbook*. 2d ed. Reston, Va.: National Council of Teachers of Mathematics, 1982.

Hirsch, Christian R., ed. *The Secondary School Mathematics Curriculum*. 1985 Yearbook of the National Council of Teachers of Mathematics. Reston, Va.: The Council, 1985.

House, Peggy A. "Alternative Education Programs for Gifted Students in Mathematics." *Mathematics Teacher* 76 (April 1983): 229–33.

―――. "Program for Able Students: District or Regional Alternatives." *Arithmetic Teacher* 28 (February 1981): 26–29.

―――. "Who Will Teach the Gifted?" *Focus on Learning Problems in Mathematics* 6 (Summer 1984): 29–38.

House, Peggy A., Markita L. Gulliver, and Susan F. Knoblauch. "On Meeting the Needs of the Mathematically Talented: A Call to Action." *Mathematics Teacher* 70 (March 1977): 222–27.

Institute for Behavioral Research in Creativity. "The Identification of Academic, Creative, and Leadership Talent from Biographical Data. Final Report." Raleigh, N.C.: State Depatment of Public Instruction, Division for Exceptional Children, 1974. (ED104039)

Jacobs, J. C. "Effectiveness of Teacher and Parent Identification of Gifted Children as a Function of School Levels." *Psychology in the Schools* 8 (1971): 140–42.

Johnson, David R., and James R. Margenau. *Mathematics Contests—a Handbook for Mathematics Educators*. Reston, Va.: National Council of Teachers of Mathematics, 1982.

Kaplan, Sandra N. *Providing Programs for the Gifted and Talented: A Handbook*. Ventura, Calif.: Office of the Ventura County Superintendent of Schools, 1974.

Kaufman, Burt, Jack Fitzgerald, and Jim Harpel. *MEGSSS in Action*. St. Louis: CEMREL, 1981.

Keating, Daniel P., ed. *Intellectual Talent: Research and Development*. Baltimore: Johns Hopkins University Press, 1976.

Kerr, Barbara A. *Smart Girls, Gifted Women*. Columbus, Ohio: Ohio Psychology Publishing Co., 1985.

Krist, Betty J. "The Gifted Math Program at SUNY at Buffalo." In *The Secondary School Mathematics Curriculum*, edited by Christian R. Hirsch. Reston, Va.: National Council of Teachers of Mathematics, 1985.

Krutetskii, V. A. "An Investigation of Mathematical Abilities in Schoolchildren." In *The Structure of Mathematical Abilities*, edited by Jeremy Kilpatrick and Izaak Wirszup. Soviet Studies in the Psychology of Learning and Teaching Mathematics, vol. 2. Stanford, Calif.: School Mathematics Study Group, 1969.

———. *The Psychology of Mathematical Abilities in Schoolchildren*. Edited by Jeremy Kilpatrick and Izaak Wirszup. Chicago: University of Chicago Press, 1976.

Lamon, William E., ed. "Educating Mathematically Gifted and Talented Children." Special issue of *Focus on Learning Problems on Mathematics* 6 (Summer 1984).

Leder, Gilah, ed. "Special Issue on Mathematically Able Students." *Educational Studies in Mathematics* 17 (August 1986).

Lenchner, George. *Creative Problem Solving in School Mathematics*. Boston: Houghton Mifflin, 1983.

Lindsey, Margaret. *Training Teachers of the Talented and Gifted*. New York: Columbia University, Teachers College Press, 1980.

Lyons, Gene. "It's Working in North Carolina." *Newsweek*, 9 May 1983, pp. 52–53.

McKim, Robert H. *Experiences in Visual Thinking*. Monterey, Calif.: Brooks/Cole, 1980.

Maker, June C. *Teaching Models in Education of the Gifted*. Rockville, Md.: Aspen Systems Corp., 1982.

Malone, Charlotte H. "Education for Parents of the Gifted." *Gifted Child Quarterly* 19 (1975).

Marland, Sidney P. *Education of the Gifted and Talented: Report to the Congress of the United States by the U.S. Commissioner of Education*. Vol. 1. Washington, D.C.: U.S. Government Printing Office, 1972.

Martin, Edward C., ed. *Elements of Mathematics*. Comprehensive School Mathematics Program. St. Louis: CEMREL, 1970–1983.

Martinson, Ruth A. *The Identification of Gifted and Talented*. Ventura, Calif.: Office of the Ventura County Superintendent of Schools, 1974.

Matros, Michael, ed. *Dialogues from the North Carolina School of Science and Mathematics*. n.p., 1985.

Mauer, Stephen B., and Walter E. Mientka. "AHSME, AIME, USAMO: The Examinations of the Committee on High School Contests." *Mathematics Teacher* 75 (October 1982): 548–57.

Mauer, Stephen B., and Anthony Ralston. "Discrete Algorithmic Mathematics." Buffalo, N.Y.: Gifted Math Program, State University of New York at Buffalo, 1984.

Minnesota Department of Education. *Minnesota Guidelines for Gifted and Talented Education*. St. Paul: Minnesota Department of Education, 1986.

Mitchell, Patty Bruce. *A Policymaker's Guide to Issues in Gifted and Talented Education*. Washington D.C.: National Association of State Boards of Education, 1981.

A Nation at Risk: The Imperative for Educational Reform. Washington, D.C.: U.S. Government Printing Office, 1983.

National Council of Teachers of Mathematics. *An Agenda for Action*. Reston, Va.: The Council, 1980.

———. *Arithmetic Teacher* 28 (February 1981).

———. *Mathematics Teacher* 76 (April 1983).

———. "A Position Statement on Vertical Acceleration." Reston, Va.: The Council, 1983.

———. "A Position Statement on Provisions for Mathematically Talented and Gifted Students." Reston, Va.: The Council, 1986.

North Carolina School of Science and Mathematics. *NCSSM Information*. 1984.

Orenstein, Allan J. "What Organizational Characteristics Are Important in Planning, Implementing and Maintaining Programs for the Gifted?" *Gifted Child Quarterly* 28 (Summer 1984): 99–105.

Passow, A. Harry, ed. *The Gifted and the Talented: Their Education and Development*. Yearbook of the National Society for the Study of Education. Chicago: University of Chicago Press, 1979.

Pegnato, C. W., and J. W. Birch. "Locating Gifted Children in Junior High Schools: A Comparison of Methods." *Exceptional Children* 25 (March 1959): 300–304.

Polya, George. *Mathematical Discovery*. 2 vols. New York: John Wiley & Sons, 1981.

Renzulli, Joseph S. *The Enrichment Triad Model: A Guide for Developing Defensible Programs for the Gifted and Talented*. Wethersfield, Conn.: Creative Learning Press, 1977.

———. *A Guidebook for Evaluating Programs for Gifted and Talented*. Ventura, Calif.: Office of the Ventura County Superintendent of Schools, 1975.

———. "What Makes Giftedness? Reexamining a Definition." *Phi Delta Kappan* 60 (November 1978): 180–84, 261.

Renzulli, Joseph S., S. M. Reis, and L. H. Smith. *The Revolving Door Identification Model*. Mansfield, Conn.: Creative Learning Press, 1981.

Renzulli, Joseph S., and L. H. Smith. "Two Approaches to Identification of Gifted Students." *Exceptional Children* 43 (May 1977): 512–18.

Richert, E. Susanne, James J. Alvino, and Rebecca M. McDonnel. *National Report on Identification: Assessment and Recommendations for Comprehensive Identification of Gifted and Talented Youth*. South Sewell, N.J.: Educational Information Resource Center, 1982.

Ridge, H. Laurence, and Joseph S. Renzulli. "Teaching Mathematics to the Talented and Gifted." In *The Mathematical Education of Exceptional Children and Youth*, edited by Vincent J. Glennon, pp. 191–266. Reston, Va.: National Council of Teachers of Mathematics, 1981.

Rising, Gerald R., and Joseph B. Harkin. *The Third "R": Mathematics Teaching for Grades K–8*. Belmont, Calif.: Wadsworth Publishing Co., 1978.

Rogers, Karen. *Review of Research on the Education of Intellectually and Academically Talented Students*. St. Paul: Minnesota Department of Education, 1986.

Ross, Arnold E. "Talent Search and Development." *Mathematical Scientist* 3 (1978): 1–7.

Sachs, Leroy, ed. *Student Merit Awards—High School*. Reston, Va.: National Council of Teachers of Mathematics, 1984.

———. *Student Merit Awards—Middle School*. Reston, Va.: National Council of Teachers of Mathematics, 1984.

Schimpfhauser, Frank. *Report of the Gifted Math Program Review Committee*. Buffalo, N.Y.: State University of New York at Buffalo, 1986.

Schlesinger, Beth M. "A Senior High School Problem-solving Lesson." In *The Agenda in Action*, edited by Gwen Shufelt, pp. 70–78. Reston, Va.: The Council, 1983.

Schwartz, Judah L., and Michael Yerushalmy. *The Geometric Supposer*. Pleasantville, N.Y.: Sunburst Communications, 1985.

"Special High Schools: A Renaissance." *Los Angeles Times*, 16 November 1983.

Stanley, Julian C., Daniel P. Keating, and Lynn H. Fox. *Mathematical Talent: Discovery, Description, and Development*. Baltimore: Johns Hopkins University Press, 1974.

Stein, S. K. *Calculus and Analytic Geometry*. 3d ed. New York: McGraw-Hill Book Co., 1983.

Stover, Donald W., Gerald R. Rising, and Eileen K. Schoaff. "Precalculus Mathematics with a Computer." Buffalo, N.Y.: Gifted Math Program, State University of New York at Buffalo, 1985.

Strang, Ruth. "Psychology of Gifted Children and Youth." In *Psychology of Exceptional Children and Youth*, edited by William M. Cruickshank. Englewood Cliffs, N.J.: Prentice-Hall, 1955.

Syracuse University. "Project Advance Handbook." Syracuse, N.Y.: Syracuse University, 1985.

Tannenbaum, Abraham J. "Post-Sputnik to Post-Watergate Concern about the Gifted." In *The Gifted and the Talented: Their Education and Development*, edited by A. Harry Passow. Chicago: University of Chicago Press, 1979.

Tanur, Judith M., Frederick Mosteller, William H. Kruskal, Richard F. Link, Richard S. Pieters, and Gerald R. Rising. *Statistics: A Guide to the Unknown*. San Francisco: Holden-Day, 1978.

Terman, Lewis M. "Intelligence and Its Measurement." *Journal of Educational Psychology* 12 (March 1921): 127–35.

Terman, Lewis M., et al. *Mental and Physical Traits of a Thousand Gifted Children*. Genetic Studies of Genius, vol. 1. Stanford, Calif.: Stanford University Press, 1926.

Terman, Lewis M., and Melita Oden. *The Gifted Child Grows Up*. Genetic Studies of Genius, vol. 4. Stanford, Calif.: Stanford University Press, 1947.

Texas Education Agency. *Myth-Conceptions about the Gifted and Talented*. Austin, Tex.: Texas Education Agency, 1979.

Tobias, Sheila. *Overcoming Math Anxiety*. New York: W. W. Norton & Co., 1978.

Toffler, Alvin. *Future Shock*. New York: Random House, 1970.

Torrance, E. Paul. *Encouraging Creativity in the Classroom*. Dubuque, Iowa: W. C. Brown, 1970.

———. *The Gifted Child in the Classroom*. New York: Macmillan Co., 1965.

Tuttle, F. B. *Gifted and Talented Students: What Research Says to the Teacher*. Washington, D.C.: National Education Association, 1978.

United States Department of Education. *Curriculum for the Gifted and Talented*. Washington, D.C.: U.S. Government Printing Office, 1978.

———. *Developing Individualized Education Programs for the Gifted and Talented*. Washington, D.C.: U.S. Government Printing Office, 1978.

———. *Finding Funds for Gifted Programs*. Washington, D.C.: U.S. Office of Education, 1978.

Van Tassel-Baska, Joyce. *An Administrator's Guide to the Education of Gifted and Talented Children*. Washington, D.C.: National Association of State Boards of Education, 1981.

Vance, James H. "The Mathematically Talented Student Revisited." *Arithmetic Teacher* 31 (September 1983): 22–25.

Wallach, Michael A. "Tests Tell Us Little about Talent." *American Scientist* 64 (1976): 57–63.

Wernick, Robert. "At Boston Latin, Time Out for a 350th Birthday." *Smithsonian* 16 (April 1985): 122–35.

Whitehead, Alfred North. Preface to *Business Adrift*, by W. B. Donhans, New York: McGraw-Hill Book Co., 1931.

Wilson, Robert C. *The Under-Educated*. Austin, Tex.: Hogg Foundation for Mental Hygiene, University of Texas at Austin, 1956.

Wolkomir, Richard. "The Winning Equation at P.S. IQ—Bright Kids, Good Teachers, Hard Work." *Smithsonian* 16 (May 1985): 80–88.

Ziv, Avner. *Counselling the Intellectually Gifted Child*. Toronto: University of Toronto, 1977.

Appendix

A Position Statement on . . .

Vertical Acceleration

THE decision to accelerate students vertically through grades K–12 is a critical one and should be made in consultation with the student, parent, teacher, counselor, administrator, and the student's academic record. Vertical acceleration should be implemented *only* for the extremely talented and productive student.

In almost all cases, a student would benefit more from a program of horizontal acceleration (enrichment) than from one of vertical acceleration. *An Agenda for Action* makes this point in stating that—

> in general, programs for the gifted student should be based on a sequential program of enrichment through ingenious problem-solving opportunities rather than through acceleration alone.

Among the topics that should be included in all mathematics programs but that also suggest important areas for an enriched curriculum are the following:

- Calculators and computers
- Problem posing and solving
- Development of facility with mathematical language
- Mental arithmetic
- Estimation and approximation
- Probability and statistics
- Intuitive geometry

An analysis of student abilities and of the mathematical applications in current careers suggests that programs of acceleration must consider the role of discrete mathematics (combinatorics, graph theory, and probability) and elementary computing skills as well as the added work of setting up and planning the calculation of answers to complex problems based on algebraic concepts and principles studied in the secondary curriculum. Stressing the development of algorithms for complex calculations can serve to tie the traditional secondary curriculum to the emerging topics listed above and help build a bridge to the collegiate study of mathematics.

The preparation of teachers of accelerated courses should be in accordance with the *Guidelines for the Preparation of Teachers of Mathematics*. These teachers should be chosen from among a district's most motivating and talented instructors. They should have the ability to challenge student thought, guide student inquiry, and make appropriate suggestions for further independent study. In addition, such teachers should be provided with support services so that the school district can maximize its gifted students' mathematical potentials.

The records of accelerated students clearly show that a significant number discontinue the study of mathematics before they complete four years of high school mathematics. Of those who do remain in mathematics until graduation, many experience extreme difficulty with the subject. Therefore, the NCTM recommends the following:

> That vertical acceleration be considered only for a limited number of highly talented and mathematically creative students whose interest and attitudes clearly

indicate that they have the ability and perseverance to complete a carefully designed sequential curriculum. For all but this select group, a strong, expanded program emphasizing mathematics enrichment is preferable.

(April 1983)

A Position Statement on . . .

Provisions for Mathematically Talented and Gifted Students

ALL students deserve the opportunity to achieve their full potential; talented and gifted students in mathematics deserve no less. It is a fundamental responsibility of all school districts to identify mathematically talented and gifted students and to design and implement programs that meet their needs. Further, it is the responsibility of mathematics educators to provide appropriate instruction for such students.

The identification of mathematically talented and gifted students should be based on multiple assessment measures and should involve teachers, counselors, administrators, and other professional staff. In determining admission to talented and gifted programs, the evaluators must consider the student's total educational development as well as his or her mathematical ability, achievement, and aspirations. Eligible students and their parents should fully understand the nature and demands of the program before making a commitment to participate. Unqualified students should not be admitted for any reason.

The needs of mathematically talented and gifted students cannot be met by programs of study that only accelerate these students through the standard school curriculum, nor can they be met by programs that allow students to terminate their study of mathematics before their graduation from high school. The curriculum should provide for all mathematically talented and gifted students every year they are in school. These students need enriched and expanded curricula that emphasize higher-order thinking skills, nontraditional topics, and applications of skills and concepts in a variety of contexts.

Therefore, the National Council of Teachers of Mathematics recommends that all mathematically talented and gifted students should be enrolled in a program that provides a broad and enriched view of mathematics in a context of higher expectation. Acceleration within such a program is recommended only for those students whose interests, attitudes, and participation clearly reflect the ability to persevere and excel throughout the entire program.

(October 1986)